Fossil Detectives

Discovering Prehistoric Britain

BBC
BOOKS

Fossil Detectives

Discovering Prehistoric Britain

Hermione Cockburn
and Douglas Palmer

This book is published to accompany the television series
Fossil Detectives, first broadcast on BBC4 in 2008.

Presenter: Hermione Cockburn
Contributors: Phil Manning, Anjana Khatwa and James Wong
Executive Producer: Fiona Pitcher
Series Producer: Kerensa Jennings

10 9 8 7 6 5 4 3 2 1

First published in 2008 by BBC Books, an imprint of Ebury Publishing.
A Random House Group Company.
Produced in association with The Open University, Walton Hall,
Milton Keynes MK7 6AA

The Random House Group Limited Reg. No. 954009
Addresses for companies within the Random House Group can be found
at www.randomhouse.co.uk

A CIP catalogue record for this book is available from the British Library.

ISBN 978 1 846 07577 3

The Random House Group Limited supports the Forest Stewardship
Council (FSC), the leading international forest certification organization.
All our titles that are printed on Greenpeace-approved, FSC-certified
paper carry the FSC logo. Our paper procurement policy can be found at
www.rbooks.co.uk/environment

Commissioning editors: Martin Redfern and Christopher Tinker
Production: David Brimble

Created by Tall Tree Ltd
Editorial: Jon Richards, Katie Dicker and Rob Scott Colson
Design: Ben Ruocco
Picture editor: Louise Thomas
Consultant: Dr Colin Scrutton

Colour origination by Dot Gradations Wickford, Essex
Printed and bound in Italy by Printer Trento Srl

To buy books from your favourite authors and register for offers, visit
www.rbooks.co.uk

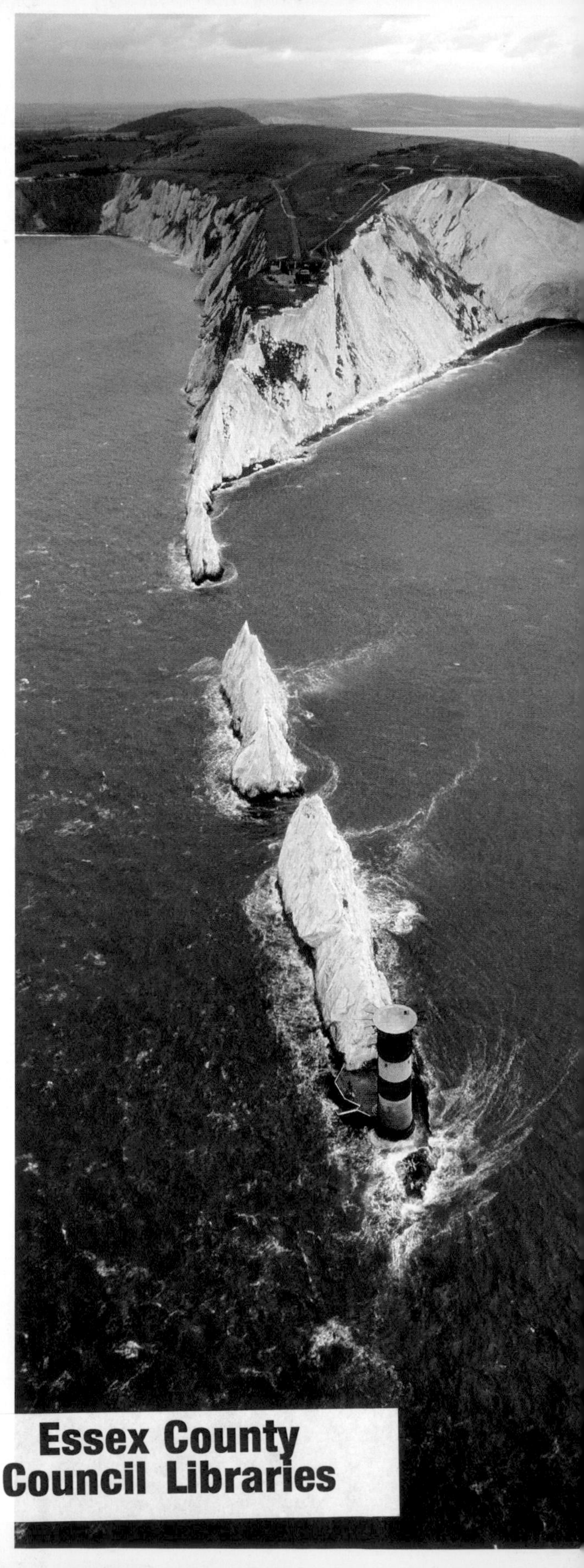

CONTENTS

Introduction

Few people forget finding their first fossil. For those of you reading this who have yet to find one, a treat awaits you. Fossils are intriguing and often beautiful objects that provide us with a tangible link to the past. When you find a fossil, you have uncovered a little bit of a mysterious lost world, often dating from many millions of years ago. Fossils are simply defined as any evidence of ancient life naturally preserved. They come in many different forms, from a coiled ammonite shell to a dinosaur footprint. Fossils are the clues we use to trace the development of life from its humble beginnings to the staggering diversity of plants and animals alive today.

By the extraordinary vagaries of geological history that have seen the British Isles cobbled together over the last billion years and more, this small region of western Europe has one of the best representations of rocks and fossils of this time span in the world. For its size and ease of access to the rocks, Britain is definitely the best location. Within a day's travel it is possible to journey through most of the geological periods that have been carved out of the last billion years. More geological

One of the earliest complete ichthyosaurs was found at Whitby in the early nineteenth century, and it can still be seen in London's Natural History Museum.

periods of time were originally defined by British geologists from British rock strata than from any other location in the world. As a result, our rocks are better known and mapped than any others. Indeed, the first modern geological map for a whole country was made in 1815, almost single-handedly, by British surveyor, canal engineer and geologist William Smith.

Despite the fact that practically every square centimetre of Britain's rocks has been hammered, measured and mapped over the last 200 years, that does not mean that nothing remains to be found. Far from it. The nature of the fossil record is such that it is always possible to turn up something new and exciting, even from sites that are still collected on an almost daily basis, such as parts of the Jurassic Coast World Heritage site in Dorset and east Devon. This is where many important first finds were made and, all the time, cliff falls and storms bring down, turn over and break up previously unseen masses of rock. This is also why such cliffs are potentially dangerous to collectors and should not be approached. There are plenty of rocks to investigate on the foreshore, however, and there is no doubt that many more new, important and interesting fossils remain to be found.

This book accompanies the television series *Fossil Detectives,* produced by the BBC's Natural History Unit in conjunction with the Open University. It owes a great debt to the production team in Bristol and all the contributors who willingly shared their expertise.

The aims of the book are similar to those of the series: to reveal some of the most important, surprising and exciting fossil stories from Britain. The book is divided into three sections: Douglas Palmer has provided a background section on palaeontology, which is followed by eight regional chapters. Hermione Cockburn has written about her three favourite stories from each region, and Douglas has added one of his. The final section includes a gazetteer of places in each region where you can find out more and try your hand at some fossil-detecting of your own. There is also some useful information on how to collect fossils and details of the major groups you are likely to encounter in British rocks. So, get ready for some time travel with the *Fossil Detectives.*

Wonderful fossils have been found along Dorset's World Heritage Jurassic Coast for centuries. In the nineteenth century the Annings sold fossils to the area's tourists.

What are fossils?

Today, most children know that fossils are things such as the shells of trilobites, ammonites and dinosaur bones. But these ancient discoveries are not the whole story – witness the recent furore about a small tubular shape found in a Martian meteorite. Was it really a fossil?

The current consensus is that the meteorite material is not a fossil, but for some time, many scientists were convinced that it was. Similar arguments have been around for centuries. This is partly because the process of fossilization can alter life forms so they take on inorganic attributes. There are also natural processes in the formation of inorganic materials that simulate organic forms. Which of the following would you consider to be a fossil, for example: a dinosaur footprint, an insect embedded in amber, a flint replica of a sea-urchin, a flint shaped like a finger, and a lump of coal?

All but the finger-shaped flint are fossils. The dinosaur footprint is known as a trace fossil and the amber insect is a body fossil. The flint replica is also a body fossil but an internal mould as well, and the coal is a chemical fossil. The finger-shaped flint is a 'sport of nature' with just an accidental resemblance to a finger. It is the kind of thing that was once regarded as a fossil, since the original meaning of the word referred to anything 'dug up'. The word 'fossil' is derived from the Latin *fossa*, meaning 'ditch'.

Historical problems

In the past, the discovery and interpretation of fossils was often very confusing and a constant source of argument. The English mediaeval chronicler Ralph of Coggeshall in Essex, recounts how, in 1171, the collapse of a river bank uncovered some huge bones

An ammonite's coiled shell was once mistaken for the remains of a snake.

that had been buried within the sediment. From his examination, Ralph concluded they were human leg bones that were so big they belonged to a giant who 'must have been fifty foot high'. We now know that the bones were almost certainly the limb bones of an extinct relative of the elephant, such as a mammoth (*Mammuthus primigenius*) or straight-tusked elephant (*Palaeoloxodon antiquus*), that lived in the region during the recent Quaternary ice ages.

While many fossils have shapes and forms that closely resemble those of living organisms, their preserving mineral material is often radically different from the substances the living organisms were originally made from. For example, the early excavation of coal often turned up long, curved 'rods' of stone with repeated scale-like patterns on the surface. Superficially they looked snake-like and were generally called 'serpent stones'. Although the shape seemed organic, the substance was clearly stony. A splendid example is preserved in the fossil cabinet of Dr John Woodward (1665–1722), who was a well-known London physician, fellow of the Royal Society and collector of natural antiquities. Woodward donated his collection and money to the University of Cambridge to found a professorship. His collection can still be seen in the University's Sedgwick Museum.

Today, we know that these Carboniferous-age fossils, called *Stigmaria*, are actually the sediment-filled internal moulds of the 'roots', or more strictly speaking underground branches, of extinct tree-sized clubmosses (lycopods) that grew to some 30 metres high in the equatorial Coal Measure forests, over 300 million years ago.

Defining a fossil today

Today, fossils are defined as the remains of any once-living organism. They are therefore the main evidence for past life, with its complex evolutionary history of origination, adaptation, speciation and extinction. Without fossils we would have no idea of the extraordinary diversity of prehistoric life and the innumerable strange groups of animals and plants that have come and gone – life forms such as the trilobites, ammonites, and graptolites as well as the dinosaurs, pterosaurs, ichthyosaurs and mammoths.

Fossil remains vary enormously, from fragments of ancient DNA to carbonized feathers and flowers, fossil tracks and footprints, but most of these fossils are exceptionally rare because they require special circumstances for their preservation. Technically, fossils include body fossils – the remains of some part of the original body tissue or skeleton; trace fossils – marks such as footprints, burrows and

The first museums were collections of weird and wonderful curiosities that often included fossils.

borings, made by organisms and preserved in the rock record; and chemical fossils – chemical residues from organisms (for example, organic-derived hydrocarbons such as coal and oil).

How to make a fossil

Fossilization is not as easy as you might think. Take a garden snail, for example. Since it is a plant-eater, there is a very remote chance that a trace fossil of this common mollusc might be preserved if a half-eaten leaf was fossilized with the telltale bite marks of the snail. When the snail dies, the soft tissue will decay or be scavenged, but the hard shell, mineralized with calcium carbonate, has a chance of surviving.

Empty snail shells are not uncommon in terrestrial soils. Even if the shells have been broken by birds, the fragments are often quite recognizable. When a snail dies, however, the original colour and surface shine of the shell is lost because the organic surface-coating

degrades and the pigments are bleached out. This makes the calcium carbonate of the shell more vulnerable to further chemical alteration, especially by acids present in rainwater or the soil. Eventually, the shell may dissolve completely. However, if the shell has been buried in a fine-grained sediment such as a mud, the sediment may form quite an accurate mould of the shell, both externally and internally. If the sediment is then lithified (turned into stone), the mould may persist and preserve sufficient detail for future fossil detectives to identify it.

Snail fossils are common but are often preserved as moulds because the calcium carbonate of their shells, and those of most molluscs, is in the unstable aragonite form, rather than the more stable calcite form, of which brachiopod and echinoderm shells are made. Most snail fossils are marine in origin, but they may be very common in certain ancient terrestrial environments, such as freshwater deposits of

The fossilized three-toed footprints of large, two-legged dinosaurs were first thought to be those of giant birds.

Even if the snail's hard shell is left intact, it has little chance of fossilization in soil.

Cenozoic age. Generally, however, terrestrial remains are less well preserved because of the constant weathering and erosion of landmasses and the recycling of their materials. The ideal circumstances for the preservation of organic remains in the fossil record are:

• a great abundance and widespread distribution of the organism;
• a longevity through geological time as a species;
• occurrence in an environment commonly preserved in the rock record where there is rapid burial and deposition, such as continental shelf seas, river deltas or lake shores;
• the presence of a tough and thick shell or skeleton made of a stable mineral such as calcite, apatite (calcium phosphate), silica or organic material such as cellulose or chitin;
• the lithification of the sediment with its organic remains to form a fossiliferous rock that will be resistant to weathering and erosion.

The most common fossils

The fossils most often seen in Britain are those of ancient shellfish such as clams (bivalve molluscs), snails (gastropod molluscs), ammonites (extinct cephalopod molluscs), trilobites (extinct arthropods), sea lilies (crinoid echinoderms) and brachiopods. (For a more complete list of common fossils see p. 158.) Strictly speaking, however, the most common fossils are not readily visible to the naked eye. They include microscopic plant pollen and spores, and the shells and skeletons of single-celled organisms. A few grammes of Cretaceous Chalk, for example, contains millions of fossil coccoliths.

Most visible fossils are marine shellfish because the vast majority of British fossil strata are the deposits of ancient shallow seas. These ancient sandstones, shales and limestones are a rock record of successive beach and sea-bed deposits that have accumulated over more than 540 million years.

Beachcombing today

Beachcombing around Britain today can yield a bounty of seashells, especially bivalved molluscs (such as the well-known cockles, mussels and razor shells), snails (such as whelks and winkles), bits of sea urchins and occasional starfish. The remains of other creatures that live within marine waters are surprisingly rare on beaches unless there has been an unusual mass mortality. This is partly because their protein-rich flesh is likely to be scavenged, and their skeletons fall apart before their bones or shells are washed up on the shore or buried in the sediment. Seaweed and other soft plant material also have little chance of preservation.

Mixtures of marine and land-derived remains are important for fossil detectives because they help to make links between the kinds of organisms that lived in these different environments. Luckily, rivers flow into coastal waters and bring with them remains of terrestrial life, ranging from whole trees and animal corpses to the abundant but microscopic spores and pollen of land plants. Estuarine and deltaic waters are important sites of significant sediment deposition, within which a great deal of organic material gets trapped. These are often important sources of hydrocarbons and fossil remains.

A beachcomber finds only a very small portion of the life that lives in and around coastal environments. Many organisms live within the sediment for protection in these hostile places. These include worms and shrimp-like crustaceans, along with many molluscs and sea urchins. The bodies of the worms and crustaceans are not normally preserved, but traces of their activity through the sediment are commonly seen as ancient burrows. In addition, shelled creatures may be fossilized in the sediment.

The remains vary from place to place, depending on factors such as the type of sediment, the presence of rocks or cliffs nearby, the depth of the water offshore, the temperature and the time of year. The size of sediment particles and the energy of the environment are important, too. Coarse sand and pebbles agitated by wind and wave currents rapidly grind down shells to such an extent that just a 'shell hash' remains. To find well-preserved fossils, it is best to target sedimentary rocks laid down in quiet waters such as muds, silts and fine-grained limestones.

Fossil-hunting

Collecting fossils from ancient marine strata is not unlike beachcombing. If you were able to wander along a Silurian beach, over 400 million years ago, you might see some familiar-looking mollusc-type shells of clams and snails. To identify them you would need the Silurian equivalent of a modern field guide. You might also see several unfamiliar shells and skeletons classified as brachiopods, trilobites, graptolites and orthoconic nautiloids.

The brachiopods could be mistaken for clam shells, but on close examination they show some important differences. One shell is bigger than the other and the bigger shell has a hole in it through which a bit of fleshy stalk still protrudes. Brachiopods are commonly known as lampshells because some look like Roman oil lamps. When alive, the animal would be anchored by the fleshy stalk to another shell, to stone or to the sea-bed sediment.

Brachiopods are one of the most common and widespread seashells of Palaeozoic times. Biologically, they are placed in their own phylum, Brachiopoda, and were filter-feeding animals that lived on the sea bed. Most were small – up to 4 cm long and, rarely, as long as 15 cm. Some 4000 fossil genera are known and there are still over 300 surviving species, some of which live in offshore waters around Britain.

Trilobites are extinct sea-dwelling arthropods, covered with a crab-like carapace on the upper surface and with numerous jointed legs and other appendages

After a storm, a beach strandline is a good place to sample the potentially preservable 'protofossil' remains of marine life.

on the lower surface. The skeleton is divided into articulating pieces that allowed the animal to roll up into a ball when threatened. Most trilobites lived on the sea bed and fed upon small organic particles from the sediment surface and seawater. Some 15,000 species are known to have lived between early Cambrian and late Permian times.

Graptolites are perhaps the strangest of the creatures to be found on a Silurian beach and may be quite difficult to spot. Looking like fret-saw blades, their remains consist of thin but quite stiff 1–3 mm-wide organic tubes with a series of tiny interconnected cups on one side. Growing to several centimetres long, their skeletons have a curious geometric, plant-like form, but graptolites are in fact colonial animals. Some grow into elegant spiral shapes with regularly spaced branches. Free-living in the water column, they drifted wherever currents took them.

The long, straight, conical shells of early Palaeozoic nautiloids were commonly washed up onto Silurian beaches. The chambered cone floated for some time after the animal died and may even have been colonized by other sea creatures before finally sinking to the sea bed or being washed ashore. The nautiloids were free-swimming animals, many of which may have lived in localized shoals. Larger forms up to a few metres long were more solitary and widely distributed. Some were so big that they preferred to live on the sea bed. In earliest Palaeozoic times, they probably had few predators except other nautiloids, at least until the jawed fish evolved in Silurian times.

Fish remains are as rare on Silurian beaches as the remains of modern fish are on today's beaches. Even if you found a Silurian fish, it might be difficult to recognize it as such because so many fish of this period were quite unlike modern jawed, bony fish. Most Silurian fish were jawless forms (agnathans) covered with a strange armour of leathery plates over the head and trunk. Only the flexible tail had more

These jawless fish of Devonian age are well preserved as fossils because of their tough body armour.

familiar scales. The fish lived on the sediment surface and fed on organic debris and bacterial mats that covered the surface. As the jawed fish evolved and diversified through Silurian times and became increasingly predatory, the jawless fish moved into fresh waters. They survived there throughout Devonian times, only to decline through the Carboniferous period as the jawed fish invaded fresh waters.

Silurian rocks in Britain

The rocks of the British landscape preserve a surprising surface outcrop of Silurian-age strata, despite the fact that the period only lasted for some 26 million years (from 443 to 417 million years ago). Some of the most extensive strata are found in the Southern Uplands of Scotland, in Cumbria and throughout the Welsh Borderlands. They are especially well exposed along Wenlock Edge in Shropshire, where there are many legally protected sites (SSSIs) and a number of Regionally Important Geological Sites (RIGS) that are more accessible to the general public. In these locations, breaking open any of the slabs of muddy limestone can reveal a glimpse of the life of the deep past. Each freshly broken slab will reveal a sight that no human eye has ever seen before. If the rock is carefully broken along a bedding plane, it will reveal an ancient sea bed and, if it has fossils on it, a moment in the geological history of the evolution of life.

Britain's life and times

A geological history

Geologists now recognize that Britain is made up of several different underlying structural units known as terranes, some of which had very different histories before being amalgamated into the British Isles. This plate tectonic history might seem strange and unbelievable to the uninitiated, but there are independent lines of evidence that support this extraordinary story of changing places on the grand scale.

The geological story can be picked up in the Precambrian eon (4560 to 542 million years ago), some 600 million years ago. At this time, southeast Britain (today's England), Wales, southeast Ireland and parts of France and Germany, were all part of a terrane known as Avalonia, which lay close to North Africa and the Gondwanan supercontinent (today's South America, Africa, Australia, Antarctica and India).

Scottish rocks record a Precambrian glaciation. Scottish tillites were the first in the world to be recognized as 'fossilized' glacial deposits of Precambrian age by James Thomson (1822–92) in 1871. The Scottish scientist James Croll (1821–90) predicted that such an ice age might have promoted the explosion of life in the later Cambrian period (542 to 488 million years ago).

From around 580 million years ago, a new ocean called Iapetus opened up and Avalonia underwent dramatic geological changes, with periods of folding and faulting and violent island-arc volcanism, like that seen in some of the volcanic islands of Indonesia, Japan and the Caribbean today. Avalonia's shallow seas were host to some of the earliest organisms, whose fossils have been preserved in the Charnwood Forest region of Leicestershire (see pp. 85–7).

Around 500 million years ago, in late Cambrian times, Avalonia moved away from Gondwana as a new ocean, called the Rheic Ocean, opened up. Over the next 60 million years, sediments and rocks were crumpled into a great fold belt, the remains of which are still familiar to us today as the Caledonian Mountains. The process joined England, Wales and

The jawless Devonian Cephalaspis had a tough, leathery headshield below which its mouth was suited to sucking up bits of organic material from the ocean sediment.

southeast Ireland to Scotland and northwest Ireland for the first time. This was the foundation of the underlying geological substructure of the British Isles, which then lay about 30 degrees south of the equator.

From an evolutionary point of view, one of the most important developments of the Palaeozoic era (542 to 251 million years ago) was fish. We now know that the ancestors of fish evolved in the Cambrian seas over 500 million years ago. These curious jawless fish (agnathans) eventually diversified into many different kinds, including some extraordinary armoured forms, by Silurian times (443 to 416 million years ago), when they also colonized fresh water. The other major innovation of Silurian times was the evolution of the vascular, or upright-growing, land plants, accompanied by the first land-going arthropod herbivores.

Late Palaeozoic times

By the beginning of Devonian times (416 to 359 million years ago), the Iapetus Ocean had closed and the British Isles, apart from the extreme south, were part of a large landmass known as Laurussia. Jawed fish had diversified to an extraordinary extent. One of the most important adaptations was that of muscular paired fins that allowed some of them to support their body weight and lift their head. These fins evolved into the first tetrapod limbs while the animals were still primarily aquatic fish-like forms. But life within the harsh environments of land required many other structural and physiological adaptations. Out of water, a fish cannot hear, see or breathe.

By mid-Devonian times, a small terrane from Gondwana, known as Armorica, had crumpled into the southern flank of Avalonia, folding the east–west trending Variscan mountains (named after the Medieval Variscia region of Germany, which is within such a mountain belt) of south Ireland and southwest England. Continued plate movements in early Carboniferous times (359 to 299 million years ago) moved Laurussia toward the equatorial zone. The sea

DID YOU KNOW? **TECTONIC PLATES**

Earth's outer layer of cool and brittle crustal rocks is broken into seven continent-sized plates and six smaller ones that have moved in relation to one another over geological time. Most plates carry both ocean and continent, with new crust being generated at constructive boundaries, known as spreading ridges, where new ocean-floor rocks form and then spread over many millions of years into new oceans.

However, since Earth does not expand, as much crust must be destroyed as is generated: this is mostly old ocean-floor rocks that are pushed back down into Earth's interior along steeply sloping subduction zones. The subduction process generates intense earthquakes and friction that melts rocks at depth, leading to volcanic eruptions at the surface.

Subduction can also lead to the collision of plates carrying continental crust and the formation of mountain belts such as the Himalayas, when the Indian continent collided with Asia some 15 million years ago.

Devonian times 416–359 million years ago

Jurassic times 200 146 million years ago

Oligocene times 34–23 million years ago

Giant dragonfly-like insects with wingspans of 70 cm lived in the coal-forming forests of Carboniferous times.

flooded over large tracts of the landscape. These warm shallow seas were ideal for reef development, and organisms included stromatolites and corals along with brachiopods, gastropods, trilobites and diverse fish including sharks. In late Carboniferous times, plate movements brought Gondwana up from the south to join Laurussia, and together they moved to the humid equator. The land re-emerged, drained by large rivers with heavily vegetated swampy floodplains, lakes and deltas. Diversification of the land plants produced vast forests and woodlands with tree-sized clubmosses, horsetails (sphenopsids) and ferns. The accumulated plant debris of these swamps and forests formed the coal seams that fuelled the Industrial Revolution.

Animal life thrived in these swampy environments, whose waters were filled with fish and newly diversifying tetrapods, including crocodile-like amphibians and the evolving egg-laying amniotes whose novel reproduction allowed them to become independent of water for the first time. The amniotes included the earliest reptiles (mostly lizard-like forms that ate small arthropods) and the abundant insects that lived in and around the lush vegetation.

As Britain and North America moved across the equator into the northern hemisphere and another

tropical semi-arid zone, other plate movements had conspired to amalgamate all the continental masses into one gigantic supercontinent known as Pangea. From space, Earth must have looked very strange, with just one landmass occupying a third of its surface. Pangean environments developed wide differences between the dry desert interior and the coast. Britain lay in the dry interior throughout Permian and into Triassic times and the beginning of the Mesozoic era (251 to 65 million years ago).

The big kill

As John Phillips showed in 1860 (see p. 31), there is a marked difference between the fossil record from the end of the Permian to that of the early Triassic period, and that is why he drew the boundary between the Palaeozoic and Mesozoic eras at this point. We now know that this is when the greatest and most puzzling extinction event in Earth's history occurred, with over 80 per cent of all life being wiped out. In the seas, the Palaeozoic trilobites, graptolites, orthoconic nautiloids and eurypterids all became extinct, along with the ancient groups of corals.

On land, there was a significant extinction of the Permian reptiles, but a more gradual turnover from the late Palaeozoic floras, with the dominant swamp plants giving way to the seed plants, cycads and conifers. With changes in their

Known as a dicynodont, this pig-sized creature was one of the few land-living creatures to survive the end Permian mass extinction.

reproductive structures, these new flora were adapting to life in drier and more upland environments.

The exact cause of the extinction is not entirely clear, and there is no evidence of a single cause. An extraterrestrial impact event has been suggested. On the other hand, there were very large-scale environmental changes in the chemistry of the oceans and atmosphere that may have been related to massive eruptions of basaltic lavas in Siberia.

Mesozoic times

Recovery in Triassic times (251 to 200 million years ago) was initially slow, with many groups never regaining their former diversity, but there were soon a number of very important evolutionary innovations. One of the surviving reptile groups acquired adaptations that were to evolve towards the mammalian condition. Another group of reptiles was to become the dominating land megafauna of the era – the dinosaurs. Some reptiles took to the seas, while yet another group took to the air as the flying pterosaurs.

The tectonic situation of the British Isles changed in Triassic times as the crustal rocks of the region began to be stretched. The underlying cause was the beginning of the break-up of the Pangean supercontinent and the initiation of the Atlantic Ocean.

In Jurassic times (200 to 146 million years ago), the seas flooded over the lowland areas of Britain. Carbonate sands and reefs developed in these warm, shallow subtropical waters, while further offshore there were deeper mud-floored basins. Marine life thrived and diversified, from the abundant, shelled, seabed-dwelling invertebrates of the shallow waters to the large predatory marine reptiles that

hunted fish, and swimming cephalopods such as the ammonites and coleoid belemnites. Lushly vegetated islands and shores were equally well populated with dinosaurs and other reptiles, alongside the small mammals that were already beginning to prosper.

The Cretaceous period (146 to 65 million years ago) is represented in Britain initially by a mixture of freshwater and shallow marine coastal environments, with nearby coastlines supplying terrestrial sediments and organic remains. Perhaps the most iconic of English rock strata is the upper Cretaceous Chalk that forms the famous white cliffs and South Downs of the south coast, with its counterpart, the North Downs on the other side of the Wealden Dome. In fact the Chalk is not particularly English. As sea levels rose in late Cretaceous times, this extensive marine deposit accumulated over much of northwest Europe.

In the waters, there were swimming ammonites, coleoid cephalopods, bony fish, sharks and the larger predatory reptiles. Particularly spectacular were the

Rhamphorhynchus was one of the many extinct pterosaur reptiles that dominated the air of Jurassic times.

giant mosasaurs that evolved rapidly from land-living lizard-like reptiles and spread around the world, all within 25 million years, only to die out during the extinction event at the end of the Cretaceous.

KT extinction

The most famous extinction event in the history of life occurred at the end of Cretaceous times when some 60 per cent or more of living organisms became extinct. This extinction is much better known than the significantly bigger end-Permian event because of the association of a significant extraterrestrial impact event with the demise of the dinosaurs and other Mesozoic reptiles.

There is no doubt that a seven-mile-wide rock from space hit the sea off the coast of Mexico, with catastrophic results for life in the region. There were also more global effects from associated changes in atmosphere and ocean chemistry but it is not yet clear whether the impact was the sole cause of extinction or exactly how it killed its victims, especially as the extinction was quite selective.

Cenozoic life

For the British Isles, the beginning of the Cenozoic era (65 million years ago to the present) is of considerable importance because of the continued opening of the North Atlantic, with extensive volcanism in the northwest. Huge volcanoes and fissure eruptions stretching from Skye south to the Giant's Causeway in Northern Ireland's County Antrim poured forth lavas as the crust split apart and we departed from North America.

Following the extinction event, a number of ecological niches that had been occupied by large reptiles were left vacant, and it took some time for them to be filled by newly evolving and diversifying mammals, birds, surviving reptiles (lizards, snakes, turtles, crocodilians) and fish (especially the sharks). Most of the early mammals were small, around a metre long, and belonged to groups that are mostly now extinct. Not until Eocene times, around 50 million years ago, did modern mammal groups begin to emerge.

Quaternary ice ages

Global cooling started around 25 million years ago, when the first ice sheets appeared in Antarctica. Yet it was not until 2.4 million years ago that the first major northern hemisphere ice sheets appeared and Earth was locked into an alternating series of cold glacial episodes, punctuated with warm interglacials. While the repeated glacials were often

Related to lizards, the mosasaurs, up to 10 metres long, were some of the most ferocious marine predators that reigned briefly in late Cretaceous times.

Small, primitive leptictid mammals may have been victims of 2-metre-tall flightless gastornithid birds of early Cenozoic times.

significantly colder on average than at any other time in the last 250 million years, the interglacials were sometimes warmer than today.

For much of the Quaternary period (1.8 million years ago to the present), Britain was essentially still part of the landmass of northwest Europe, connected by a landbridge called Doggerland, until huge floods of meltwater from northern ice sheets cut the straits of Dover. With ice blocking the northern North Sea, meltwater ponded in the southern part of the North Sea and eventually flooded over the landbridge into the English Channel. Such was the energy of the flood waters that they eroded a permanent channel. Repeated upland glaciation of Britain not only removed just about everything from the landscape down to the bare rock but also eroded huge amounts of rock. The debris was spread down glaciated valleys and out over the surrounding lowlands.

In our anthropocentric view of past life, the record of our extinct ancestors and relatives is of particular interest. There have been significant new discoveries that are revolutionizing our view of early British colonization. Evidence of human-related activity (probably *Homo heidelbergensis*) dating back some 700,000 years has recently been recovered from Pakefield in East Anglia (see p. 153), for example. Very fragmentary human-related remains, dating back some 400,000 years, were first found at Swanscombe in Kent by a dentist, Alvin Marston, in 1933. They are now thought to belong to early *Homo neanderthalensis*. More recent discoveries include hand axes dredged offshore from East Anglia. They show that these

people were occupying the Doggerland connection to mainland Europe around 100,000 years ago. Even 60,000 years ago, the Neanderthals seem to have hung on. The first evidence for our immediate human ancestors, *Homo sapiens*, in Britain comes from cave finds such as the buried skeleton excavated by William Buckland in South Wales and dates back to around 27,000 years ago (see pp. 110–11).

The future

Geologically, Britain is remarkably stable. It rarely suffers from any significant earthquake activity, and there is no current risk of volcanic activity. However, we should not be complacent. Since the end of the last glaciation, crustal rocks have been slowly 'rebounding' and rising in the north of Britain but not in the south. This, coupled with global sea-level rise related to climate warming and instability, is not good news for low-lying coastal areas in the south. They will become more prone to flooding as the decades pass.

Before its extinction, the spectacular giant deer Megaloceros, *with its 2-metre-wide antlers, was pictured in early cave paintings.*

What's in a name?

Classifying fossils

Common names for extinct fossil groups, such as dinosaurs, trilobites and ammonites, are very familiar these days. Fossils also need to be classified scientifically according to the same system as living species.

Biologists classify all living things in a hierarchy of groups, known as taxonomic ranks, according to their shared characteristics. The top rank, 'kingdom', broadly groups organisms according to whether they are plants, animals or unicellular life forms, such as bacteria. Below this comes a rank called 'phylum', which groups, for instance, animals according to the general way in which their bodies are organized. Each phylum then contains several 'classes' of organism with more specific traits in common. For example, clams, mussels, scallops and oysters are all members of a class called Bivalvia – water-dwelling molluscs characterized by two hinged shells enclosing the soft body parts. Members of the Class Bivalvia (with some 12,000 living species) have biological features that ally them with other classes of animals, such as snails and slugs (Class Gastropoda – 103,000 living species); squids, cuttlefish, octopus and nautiloids, plus the extinct ammonites (Class Cephalopoda – 730 living species); and several other minor classes that are generally less familiar such as the tusk shells (Class Scaphopoda – 400 species) and chitons (Class Polyplacophora – 900 living species). Altogether, these classes are grouped within the large and ancient Phylum Mollusca, which contains over 117,000 living species and many thousands of fossil species that range back over geological time to the early Cambrian, over 530 million years ago.

Limpets, with their cap-shaped shells, and shell-less slugs are both members of the Class Gastropoda within the Phylum Mollusca, but only the shelled forms are normally fossilized.

Familiar species of bivalve molluscs, such as the oysters, cockles, mussels and razor shells that can easily be found on our beaches, also have their own scientific names, such as *Ostrea edulis* (oyster), *Cerastoderma edule* (cockle) and *Ensis arcuatus* (razor shell). Common names such as cockle and mussel are used in everyday speech, and are far less cumbersome than the scientific names, but few fossils have common names aside from the general category names such as trilobite, ammonite, brachiopod and so on. The fossil detective needs to be very particular in identifying what has been found, and that means using scientific taxonomy, just as any serious gardener or bird-watcher has to do.

Genus and species

We humans belong to the genus *Homo* (Latin for 'man') and the species *sapiens* (Latin for 'wise'). These are the lowest two taxonomic ranks. It is normal to print these generic and specific names together in italics. This Latin 'binomial', as it is called, was formulated by eighteenth-century Swedish botanist Carolus Linnaeus (1707–78). As European explorers brought back more and more unknown plants and animals from around the world, they had to find some systematic way of identifying and classifying them. Linnaeus was one of the first scientists to try to systematize a growing confusion of vernacular and Latin names.

This scientific process requires any potentially new form to be compared with already known organisms to check whether it really is new to science.

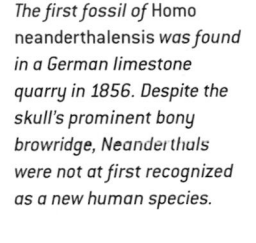

The first fossil of Homo neanderthalensis *was found in a German limestone quarry in 1856. Despite the skull's prominent bony browridge, Neanderthals were not at first recognized as a new human species.*

Generally, new fossil finds have some level of similarity with some known organism or other. Completely novel and unique organisms are very rare. For instance, most helically coiled shells belong to gastropod molluscs (snails) and so can be compared with other shells of this class of animal, no matter where they have come from geographically or in what age of rock they were found. But even here there are exceptions and, to confuse the unwary even more, there are plenty of snails with flat, spiral shells and gastropods whose shells do not evidently coil at all, such as the common limpets. Slugs are also gastropods and, since they do not have shells, they are not represented in the fossil record.

So, similarities between organisms can be deceptive, and the fossil detective has to make a thorough investigation and be wary of false clues. Linnaeus himself made plenty of mistakes and grouped together many organisms that we now understand have no close biological relationships. Superficial similarity can be particularly problematic for fossil detectives who do not have complete living organisms to compare. Fossil bones are often broken when found and are difficult to identify. The correct identification of many of the extinct fossil reptiles such as the dinosaurs was beset with these problems when they were first found in the early decades of the nineteenth century.

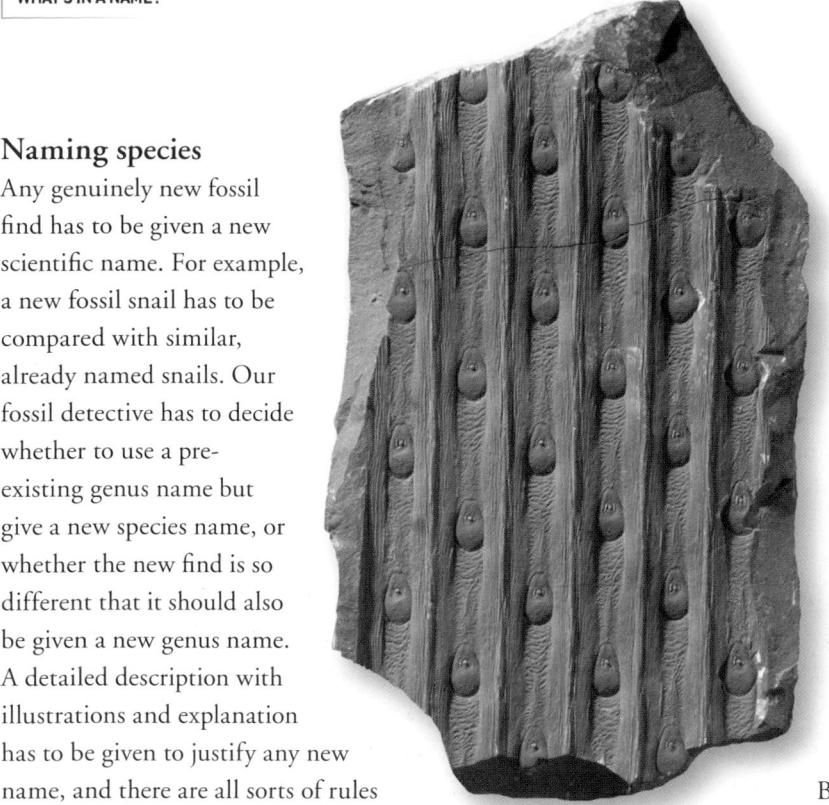

Naming species

Any genuinely new fossil find has to be given a new scientific name. For example, a new fossil snail has to be compared with similar, already named snails. Our fossil detective has to decide whether to use a pre-existing genus name but give a new species name, or whether the new find is so different that it should also be given a new genus name. A detailed description with illustrations and explanation has to be given to justify any new name, and there are all sorts of rules that have to be followed (published by the International Commission on Zoological Nomenclature for animals and by the International Commission for Botanical Nomenclature for plants).

The results then have to be submitted to a recognized scientific journal which, if it accepts them, will send the manuscript to a known expert for what is called 'peer review' to check whether the claims made stand up to rigorous criticism. If they do, the paper will be duly published and the new species will join the growing ranks of known taxa. Since over 103,000 living species of snails are already known, along with many thousands of fossil species, the chances of finding a new one are slim but not impossible. However, the task of checking the records of those already known is so great that it is best left to a specialist. If you really want to find a fossil that is new to science, you are much better off with the dinosaurs, since there are only some 600 or so genera known. The only problem is to first find your dinosaur or at least the critical diagnostic bits, such as the skull.

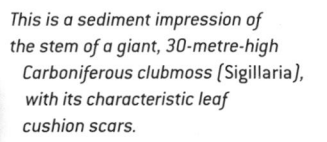

This is a sediment impression of the stem of a giant, 30-metre-high Carboniferous clubmoss (Sigillaria), with its characteristic leaf cushion scars.

What is a species?

For Linnaeus and his contemporaries, there was a basic understanding of what a biological species is – namely, a group of like individuals that can interbreed and produce viable young that can, in turn, interbreed. Indeed, the scientists of his time thought that all species were fixed or 'immutable', as they called it. But Linnaeus also recognized that, while species can be differentiated, some have sufficient similarity for them to be placed in the same genus.

For instance, today we recognize that our species is just one of several that belong in the genus *Homo*. The first other member was *Homo neanderthalensis*, originally distinguished in 1863 by an English palaeontologist, William King (1809–86), who was professor of geology at Queen's College, Galway (now the University of Galway). Today our genus includes at least eight species, and we know that there is an evolutionary connection between them, although this was not known in Linnaeus's time.

Indeed, Linnaeus got into trouble when he first named humans as *Homo sapiens* in 1758, because he grouped us with the apes and monkeys in his Order Anthropomorpha (later changed to Order Primates), which was one of several orders within the Class Mammalia. Linneaus's critics did not like the fact that he made the association but he challenged them to show what substantial anatomical differences exist between monkeys, apes and humans. Classification for Linnaeus was mainly a hierarchical method of sophisticated 'pigeon-holing' into species, genus, order, class and kingdom, and certainly did not have any evolutionary connotation. Importantly, Linnaeus believed that the higher rankings of order and class reflected natural groupings of life. Overall, he made his intention quite clear by quoting from Psalms in the Old Testament: 'O Jehovah! How manifold are thy works! In wisdom hast thou made them all; the earth is full of thy riches ...' (Psalms, chapter 104, verse 24).

Within Linnaeus's lifetime, his book *Systema Naturae*, first published in 1735, went through more than ten editions, and the number of known species described within it grew from a few hundred to include some 7700 species of plants and 4200 animals. Well over 1.5 million species of living organisms have now been described, and it is thought that there are between 5 and 10 million living species in total.

Numbers of living species

A recent estimate of the different kinds of living species breaks down a total of some 1,587,000 into protists (30,000), monerans (4800), viruses (1000), fungi (69,000), algae (27,000), vascular plants (248,400), chordates (43,000), non-insectan arthropods (123,000), insects (925,000) and other animals (116,000). Arthropods are by far the most common species today and by far the majority of them are insects. To give a comparative measure, there are only some 4500 living species of mammals. An important consideration here is which, of all these groups, are actually represented in the fossil record?

What is a fossil species?

It is very important to realize the difference between biological and fossil species. The ultimate breeding test cannot be applied to fossil species and essentially they are what is known as 'form species', since their identity is dependent upon the preservable characteristics of their hard parts rather than any soft tissue or genetic identifiers.

There are also trace fossils, which have their own taxa because so few of them can be matched to the organisms that made them. Such fossils are very important for informing us about behaviour. For instance, analysis of dinosaur footprints can give an indication of how fast the animals could move. The 3.5-million-year-old human-like footprints at Laetoli

The pattern of leaf cushion scars identifies this plant fossil as Lepidodendron, a 40-metre-high clubmoss.

in Tanzania tell us that even the little australopithecine ape-men could walk upright, more like modern humans than apes.

Some organisms, especially plants, tend to disintegrate into their component parts after death, so much so that different species names have to be given to these parts. For instance, many of the large trees that grew in Carboniferous-age Coal Measure forests are known by separate names for their root systems, trunks, foliage and reproductive parts. A well-known and common tree-sized clubmoss has rooting structures known as *Stigmaria*, a trunk with leaf scars called *Lepidodendron*, foliage known as *Cyperites* and reproductive structures called *Flemingites*. Plant spores and pollen are also given entirely separate names because it is very rare that they can be matched exactly with the parent plant, unless they are relatively modern fossils that can be matched with living species.

However, palaeontologists feel reasonably confident that the methods used to distinguish fossil species produce results that are close to those used by biologists for living species. It does seem that hard parts of living species, which would provide potential fossils, also conserve characteristics that allow a comparable species level of distinction. Interestingly, the fossil-based distinction of Neanderthals as separate from *Homo sapiens*, originally based on certain skull characteristics, was often criticized by biologists, who thought it more likely that the Neanderthals were, at best, a subspecies of *Homo sapiens – Homo sapiens neanderthalensis*. However, the recent recovery of fossil DNA from Neanderthal remains seems to support a distinction at the species level. This in turn is supported by archaeological evidence, which indicates that the two species diverged from a common ancestor over 300,000 years ago.

Although insects are among the most common creatures on Earth, their fossil remains are rare, except in certain environments such as ancient lake sediments.

How many fossil species are there?

Surprisingly perhaps, we do not know exactly how many fossil species there are, but scientists estimate that in the region of some 250,000 have been described, covering the whole period during which life has existed. This cannot be anywhere near the true number, however. Despite the fact that fossil species have been named and described for over 300 years, the palaeontological community is much smaller than the biological one. Also, crucially, there is an inevitable bias in the fossil record towards those organisms that have preservable hard parts.

As we have seen, almost three-quarters of all living organisms today – well over a million species – are arthropods or plants. Yet the total number of plant and arthropod fossil species is perhaps ten thousand or so, and that is for all time. This is largely because many terrestrial organisms and sediments are not preserved as fossils and rocks. The best preserved of the arthropods are the aquatic groups such as the trilobites and ostracodes. Although incredibly diverse and common today, terrestrial arthropods such as the insects are only fossilized in certain environments such as low-energy lake and estuarine deposits.

For all the vast abundance of land plants, their fossil record is extremely fragmentary, with reproductive structures separated from leaves and branches, stems and roots. Tough leaves, pollen, seeds and woody stems can only be preserved in different sedimentary environments. Curiously, coal, which is made of plant debris, is not as good as you might think for preserving plant fossils. The process of coalification tends to destroy identifiable plant structures. However, strata adjacent to coal seams do often preserve plant fossils.

Most organisms without shells or skeletons of some sort, such as the abundant fungi, many of the protists and worms, are barely represented in the fossil record. Few of the countless other soft-bodied invertebrates are preserved, and nor are many of the plants that do not have woody tissues, although their spores and pollen are better represented.

Making an estimate

It is possible to do some very crude calculations of the true potential numbers of fossil species. Supposing life today does consist of 5 million species and abundant life has been around for at least 500 million years, we could estimate the total number of fossil species if we knew the average duration of a species. Taking a very conservative longevity estimate of 5 million years, it would seem that there might have been a turnover of around 100 species 'generations' over the last 500 million years. Since life on land did not really get going until about 350 million years ago, such a crude extrapolation is not reasonable. But if we just consider the past 300 million years alone, with 60 species 'generations', 60 times a total diversity of 5 million species gives a total species number of 300 million potential fossil species, including

soft-bodied species that we are unlikely to find in the fossil record.

Even if the fossil species with preservable hard parts were only one tenth of that total, there could be something like 30 million fossil species. There are an awful lot of 'ifs' and 'buts' here. Nevertheless, our known fossil record is probably only about 1 per cent of the potential fossil record, which leaves an awful lot out there waiting to be identified. If you have a burning desire to discover new life forms, become a fossil detective and, with some good luck, a few years of hard searching in the rocks and a few more in the literature, there is a good chance that you can put your name to a new fossil species.

Classification and evolution

As the number of described organisms increased exponentially from the late eighteenth century into the nineteenth century, their hierarchical classification became increasingly elaborate, with the introduction of new categories of family and order.

Plant tissues range from delicate structures such as flowers to tough wood. Wood is more often preserved, but it can be difficult to identify unless it is very well preserved, like this cedar fossil.

Now, the nested scheme grows from species to genus, then family, order, class, phylum and finally, at the top of the pyramid, kingdom. Increasingly the scheme was seen to have evolutionary implications as Darwinian evolutionary theory developed. By the end of the nineteenth century it was widely accepted that life, including fossils, could be classified and represented as an evolutionary tree rooted in a primitive unicellular past and evolving through time, with invertebrate creatures evolving into increasingly more complex backboned animals, and mammals and humankind seen to be crowning the whole edifice. For backboned animals, the evolutionary sequence was seen as developing over time from fish through amphibians into reptiles, birds and mammals.

As Darwin recognized, there was a major problem for evolutionary theory in the absence of evidence among living animals for the branching points between class-level categories. The expectation was that the fossil record should be able to provide the missing evidence, but even by the mid-nineteenth century, it had not been found. Darwin had a good idea as to why this was so, and devoted a chapter of his groundbreaking 1859 book *On the Origin of Species by Means of Natural Selection* to the question of 'The Imperfection of the Geological Record'. By a stroke of luck, the first good 'missing link' from the fossil record was found within a few years.

A critical fossil link

In 1861, fossils of an early primitive 'bird', called *Archaeopteryx*, were discovered in late Jurassic strata in Germany. The fossils revealed a mixture of bird-like and reptilian features, and clearly demonstrated an evolutionary link between reptiles and birds. Detailed analysis also showed that the structure of the pelvis and feet of birds showed an evolutionary relationship with archosaur reptiles,

This asymmetric flight feather belongs to Archaeopteryx, the oldest fossil bird, from late Jurassic times.

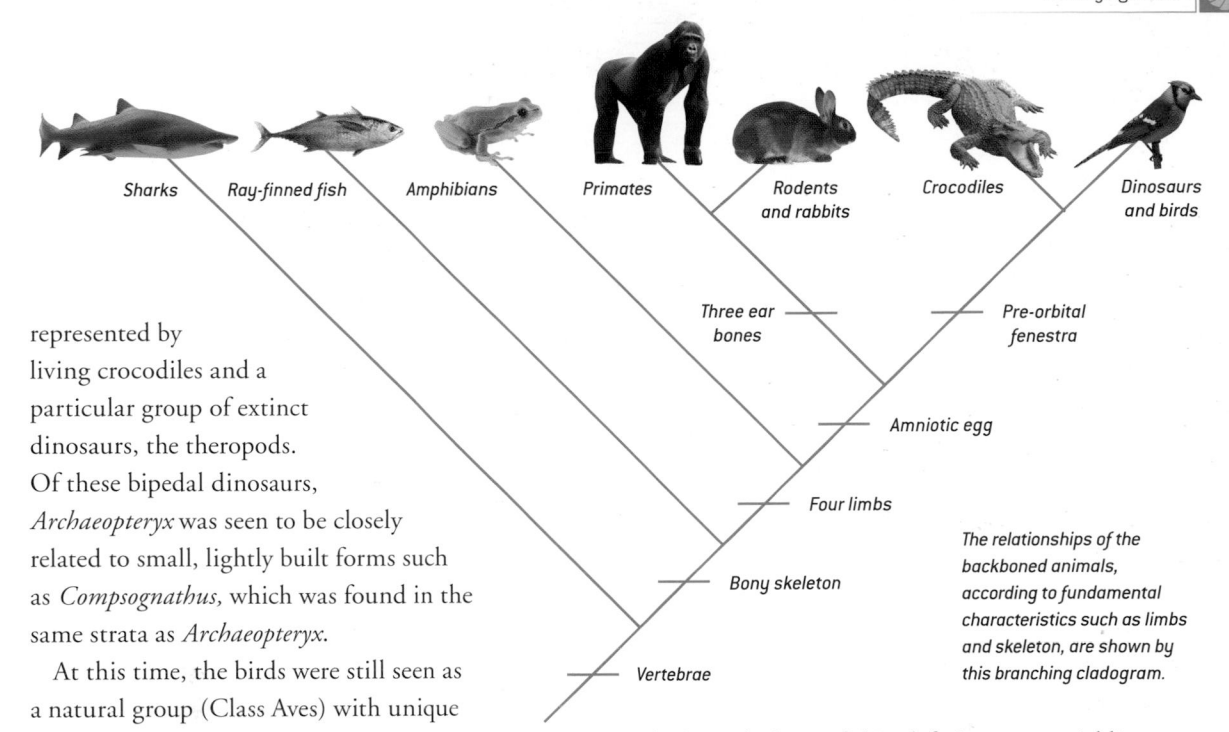

Sharks | Ray-finned fish | Amphibians | Primates | Rodents and rabbits | Crocodiles | Dinosaurs and birds

Three ear bones

Pre-orbital fenestra

Amniotic egg

Four limbs

Bony skeleton

Vertebrae

The relationships of the backboned animals, according to fundamental characteristics such as limbs and skeleton, are shown by this branching cladogram.

represented by living crocodiles and a particular group of extinct dinosaurs, the theropods. Of these bipedal dinosaurs, *Archaeopteryx* was seen to be closely related to small, lightly built forms such as *Compsognathus,* which was found in the same strata as *Archaeopteryx.*

At this time, the birds were still seen as a natural group (Class Aves) with unique features such as feathers and a basic flight capability using adapted forearms covered with asymmetrical flight feathers. The recent discovery of feathered dinosaurs without the ability to fly shows that feathers can no longer be seen as a feature unique to birds, and that they originally evolved for other purposes such as insulation or display. Birds are, in effect, a group of theropods best referred to as avialean dinosaurs. Again, it is the fossil record that has provided this evolutionary insight.

Problems with classification

Many scientists anticipated that recognition of the evolutionary significance of certain characteristics combined with the application of Linnaean classification would allow the establishment of the evolutionary relationships between many organisms. But it also became apparent that for many groups, especially fossil ones, it would be virtually impossible to realize these relationships on anatomical grounds alone. Factors such as geological age range, appearance and disappearance in the rock record had also to be taken into account. Many shell fossils are identified largely on the basis of shell ornamentation,

which might be useful in defining sequential lineages of species but is of little or no use in establishing evolutionary affinities with others that belong to the same overall group. Even within living animals such as the mammals, the evolutionary inter-relationships and ancestry of fairly high-category groups such as the tree shrews (scandentians), flying lemurs (dermopterans) and bats (chiropterans) were problematic, and their fossil record was too sparse to be of much help.

Cladistics

It was the huge problem of classifying the extraordinary abundance of insects that really highlighted the problems with Linnaean classification. A German entomologist, Willi Hennig (1913–76), instigated a conceptual revolution with the first method for establishing a true natural classification or phylogeny. In Hennig's system, the history of the descent of living beings is based on 'monophyletic' groups that include an ancestor and all of its descendants. These relationships are represented by branching diagrams called cladograms. Hennig's 'phylogenetic' method is called cladistics and is increasingly replacing Linnaean classification.

What is the rock and fossil record?

Fossil remains are scattered throughout the rock strata that underlie most of the landscapes of Britain. The rocks form what is known as the stratigraphic record, while organic remains make up the fossil record.

The discovery and mapping of this rock and fossil record over the past 200 years has revealed a complex history of environmental and evolutionary change throughout geological time. Recognition of the sequence of rock strata and its subdivision on the basis of fossils led to the establishment of successive periods of geological time.

Dividing time and mapping rocks

Although initial attempts to subdivide the deep history of geological time were made several centuries ago, the modern scheme was largely devised during the nineteenth century. As geologists made the first pioneering attempts at mapping the distribution of rocks and their three-dimensional relationships, they began to build up generalized successions of strata in the form of vertical columns with the oldest strata at the bottom and the youngest at the top. These stratigraphical columns acted as chronological keys to the sequence of depositional events.

While mapping the outcrops of different rock types across the landscape, geologists recorded how the strata were inclined, in terms of dip relative to the horizontal and direction relative to north. Geological structures such as folds and faults were also depicted, and vertical sections were constructed to show how layers of strata related to one another across the map.

The best-known pioneering geological map-maker in Britain was William Smith (1769–1839). Using his experience as an engineer and surveyor, Smith discovered that particular groups of strata, such as the Chalk and the Coal Measures, could be identified by their characteristic strata and by the particular fossils they contained. He collected samples of the rocks and fossils and arranged them in the stratigraphic order in which they occured. By 1815, Smith had completed his geological mapping of England and Wales and published the first detailed compilation map of the geology of the whole region. In 1817, he published a detailed vertical 'cross-section' from London to North Wales, in which he recognized some 33 stratigraphic units from the youngest 'sand and brickearth' of the London Basin down to the oldest 'Killas slate' in Wales.

Smith's mapping of British strata preceded the recognition of today's formal period, epoch and age names. The outcrops of strata that he recognized and mapped were those whose names had been used by miners and quarrymen for centuries. Chalk was one of them, as was the Coal Measures, for obvious reasons.

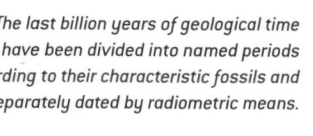

The last billion years of geological time have been divided into named periods according to their characteristic fossils and separately dated by radiometric means.

Millions of years ago

		Millions of years ago
CENOZOIC	Quaternary	0
		1.8
	Neogene	
		23
	Palaeogene	
		65
MESOZOIC	Cretaceous	
		146
	Jurassic	
		200
	Triassic	
		251
PALAEOZOIC	Permian	
		299
	Carboniferous	
		359
	Devonian	
		416
	Silurian	
		443
	Ordovician	
		488
	Cambrian	
		542
LATE PROTEROZOIC	Ediacaran	
		630
	Cryogenian	
		850
	Tonian	
		1000

Part of William Smith's beautiful and pioneering geological map of 1815 shows the distribution of strata around the Bristol region.

Geological time

Most of the geological periods recognized today were formally named in the first few decades of the nineteenth century, when regional and national geological mapping was taking place for the first time. For example, although not formally named until later in the year, the Cretaceous period was first recognized in 1822 by an aristocratic Belgian, Jean Baptiste Julien D'Omalius d'Halloy (1783–1875). D'Omalius d'Halloy called the white chalk strata the Terrain Cretacé (derived from the French 'craie', meaning chalk). Later in 1822, the Terrain Cretacé was anglicized to Cretaceous by the British geologists William Conybeare and William Phillips. Today the Cretaceous period is recognized worldwide as a unit of geological history and time.

By the early decades of the nineteenth century, it was recognized that geological mapping was an important economic tool for an emerging manufacturing nation such as Britain that relied on natural resources to fuel and supply the Industrial Revolution. There was an increasing demand for coal, iron and other metal ores, as well as clay, limestone and other stone for building and road metal. The quasi-military Ordnance Survey, founded in 1797, was ideally positioned to extend its role in mapping and evaluating land, but to begin with, this was a slow and gradual process. It was only in 1835 that an independent Geological Survey finally emerged under the leadership of Henry De la Beche (1796–1855).

In Britain and abroad, this process of formal geological mapping and description of rocks,

---

BIOGRAPHY: **WILLIAM SMITH**

In 1819, the Geological Society of London compiled a geological map of Britain that used much of Smith's information without giving him the credit. The map's publication undercut sales of Smith's own map, which was a hugely expensive enterprise. Smith ran into serious financial problems, alongside some injudicious investments, and was declared bankrupt and sent to King's Bench debtors' prison in Southwark, London. Smith was ruined. On his release, it was only the intervention of a one-time employer and kindly benefactor, Sir John Johnstone MP and fellow of the Geological Society, that Smith was able to return to work as a surveyor and land agent in the north of England.

Smith's geological rehabilitation began in the 1830s with the publication of the first volume of Charles Lyell's *Principles of Geology*. Lyell praised both Smith's 1790 *Tabular View of British Strata*, in which he formulated 'the laws of superposition of stratified rocks; that ... they might be identified at very distant points by their peculiar organized fossils', and his 1815 map.

Effectively, Lyell admitted the plagiarism of Smith's map by the gentlemen of the Geological Society. He went on to describe Smith's map as 'a lasting monument of original talent and extraordinary perseverance'. Such praise prompted the Society to restore Smith's reputation, and in 1831 he was awarded the new, prestigious Wollaston Gold Medal. The president of the Society, Adam Sedgwick, Woodwardian Professor at the University of Cambridge, felt 'compelled ... to perform this act of filial duty ... and to place our first honour on the brow of The Father of English Geology'. The title has stuck ever since.

In the early nineteenth century, Cambridge academic Adam Sedgwick was the first to recognize that these ancient Welsh slates belonged to a distinct period of geological time, which he named the Cambrian.

along with any fossils or minerals they contained, helped drive the need to standardize the naming of identifiable and mappable sequences of strata. It was a highly competitive endeavour, in which British geologists acquitted themselves very well. Charles Lyell (1797–1875) named the Pleistocene, Pliocene, Miocene and Eocene in 1833. Adam Sedgwick (1785–1873) and Roderick Murchison (1792–1871) named the Cambrian in 1835, and the Devonian in 1839. Murchison had also named the Silurian in 1835 and the Permian in 1841. Finally, in 1879, Charles Lapworth (1842–1920) carved out the Ordovician period from overlapping Cambrian and Silurian strata, in order to resolve a bitter dispute that had developed between Sedgwick and Murchison over the boundary between the two periods (see below).

The stratigraphic column that we are familiar with today has taken the professional lifetimes of countless geologists to compile. Essentially, it records the chronological succession of sedimentary strata and their division into a hierarchical series of time-related rock units. The purpose of all this technical stuff is so that geologists from different countries can use the same generally accepted stratigraphic names when they talk to one another and discuss geological history and the evolution of life. For instance, the Cambrian sediments laid down in North Wales, where the system was first named by Sedgwick in 1835, are now known to belong to the same

geological period as the famous fossiliferous Chengjiang deposits of south China and the Burgess Shale of the Canadian Rocky Mountain state of British Columbia. When Chinese, North American and British geologists refer to strata within the Cambrian period, they all know that they are talking about the same interval of geological time.

A much more familiar episode in geological history is that of Jurassic times. Thanks to Steven Spielberg's 1993 film *Jurassic Park*, based upon the novel by Michael Crichton, the name Jurassic is familiar, but how many know the story behind the name? Since the advent of radiometric dating, which measures the decay of radioactive isotopes, this period is now known to have extended from 200 until 146 million years ago and is particularly renowned for its fossil dinosaurs. The period was first recognized in the early nineteenth century and was based upon the distinctive limestones of the Jura Mountains that flank the European Alps. The Jurassic was formally established in 1839 by the German geologist Baron Christian Leopold von Buch (1774–1853), who also compiled the first geological map of Germany in 1832.

Periods, epochs and ages

Geologists went on to subdivide the Phanerozoic periods into epochs (generally early, middle and late) with durations of some tens of millions of years. These in turn are split into ages. The global geological community has gradually been establishing internationally recognized standard names for the equivalent rock units, with chronological dates and reference points for their boundaries.

One of the great spin-offs from the fossil collecting that was going on during the nineteenth century was an ever-increasing and improving measure of changes in the succession of life over time. Clearly different groups of fossils originated at different times in the past, flourished and then decreased and sometimes died out altogether. Patterns of change began to emerge, and questions arose as to the possible meaning of such patterns. By 1860, John Phillips (1800–74), Oxford professor and nephew of William Smith, was able to construct a general view of the changing diversity of life over geological time. In this he showed how the diversity of life has increased over time but was also punctuated by two great crises. The first occurred at the end of the Permian and the second at the end of the Cretaceous. These events separated the history of life into three great eras: the Palaeozoic (meaning 'ancient life'), Mesozoic ('middle life') and Cenozoic ('recent life').

Over the years since Smith's work, knowledge of the fossil record has been revolutionized thanks to the efforts of countless fossil detectives, both professional and amateur. We now have a huge data bank on the diversity of past life and its distribution through time and space. Some fossil distribution patterns were very perplexing and required fanciful explanations, such as land bridges across oceans, but the plate tectonic revolution of the last few decades has resolved many of these problems. We now know that, throughout geological history, continents have moved over Earth's surface as oceans have opened and closed, driven by changing patterns of heat flow from Earth's core.

A trilobite from the Burgess Shale World Heritage Site in Canada's British Columbia identifies these strata as belonging to the Cambrian period of geological time.

From several lines of independent evidence, geologists have been able to reconstruct a pretty good picture of plate movement over the last 700 million years and form some speculative ideas about what was happening before that. The most reliable information is the most recent, and when we run the 'tape' backwards from the present, an extraordinary picture emerges, with some continents and continental fragments moving over thousands of miles. One of the most spectacular movements was that of India, which broke away from Africa in Cretaceous times and crossed from the southern hemisphere into the northern hemisphere to eventually collide with Asia some 20 million years ago, generating the Himalayas and the high Tibetan Plateau.

Britain's geological history may not have produced such dramatic results, but in some ways it is even more extraordinary. It is a story of drastically changing environments as the jigsaw of continental pieces came together to form the geological entity that we recognize as Britain today (see pp. 14–19). These environmental changes formed the background to the evolution of life and resolve many of the puzzles that confronted the nineteenth-century pioneers of British geology and palaeontology.

What use are fossils?

Fossils have been put to numerous uses, from religious to scientific, since our extinct relatives, such as the Neanderthal people, first made collections.

For the Neanderthals, fossils may simply have been objects of curiosity or were perhaps imbued with some magical properties. In the days of the ancient Greeks, some fossils were venerated and placed in temples because they were taken as evidence for the existence of the giants of myths and legends. The first systematic collection of fossils in Renaissance times was almost purely as objects of curiosity and debate rather than for any utilitarian purpose. However, good specimens soon became valuable and were exchanged for money, because most of the people who wanted them for their collections would not 'demean' themselves by going out and actually collecting their own fossils.

So the value of fossils was primarily intellectual, aesthetic and commercial, and aimed at satisfying the demands of the gentlemen collectors. But with the scientific and industrial advances of the late eighteenth century, the use of fossils began to change quite radically. By this time it was generally accepted that fossils were the remains of past life entombed in successive layers of sedimentary rock, but the age of Earth's rock strata and their contained fossils was still a highly contentious issue because of the prevailing culture in Europe and the extended Christian world. The Judeo-Christian Old Testament version of Creation was taken as a historical document of fact. Calculations based on Old Testament genealogies estimated that Earth was no older than some

Over the 200 years since the Industrial Revolution began, much of Britain's Carboniferous-age coal has been burned up, fuelling global warming.

Shark's teeth were among the fossils used to characterize the Chalk strata of the Cretaceous period in the early nineteenth century.

6000 years. Such was the power and influence of this view that many scientists who were encountering evidence that contradicted this dogma were reluctant to publish their findings.

The geological revolution

By the end of the eighteenth century, however, the cultural climate had changed sufficiently to allow increasingly open criticism of this religion-based constraint on geological time and the Earth's age. On a more practical level, a growing number of geologists just avoided the problem and described what they found. As the Industrial Revolution required more and more raw materials, including rocks and minerals from limestone to iron ore and coal, it became increasingly important to know where there might be untapped reserves, especially of underground coal. Huge sums of money were wasted by landowners sinking expensive shafts through the wrong kind of rock strata in the search for coal. At the time there was also no reliable and detailed map showing the distribution and structure of geological strata of different ages. The revolution of the late eighteenth to early nineteenth century in the mapping and classification of the geological record helped to address this problem. Much of this new mapping system was based on work in Britain, as described on pp. 28–9.

Biostratigraphy

By the mid-nineteenth century and the establishment of the Geological Survey of Great Britain, the use of fossils in mapping, and for the stratigraphic division of geological time, was normal practice and became known as biostratigraphy. Inevitably, large numbers of fossils, many of them new to science, were discovered by the survey map-makers, and these fossils had to be identified. Palaeontologists and specialist collectors were employed to describe, illustrate and catalogue the growing reference collections. There were questions about their classification and distribution, both geographically and chronologically. It soon became evident that some fossils were of much more use than others for the fine subdivision of strata and their correlation between outcrops across the country and further afield.

On the one hand it was evident that there was an overall pattern of fossil groups that appeared and disappeared at different times in the general succession of strata. Extinct invertebrates such as trilobites and graptolites were found mainly in early Palaeozoic strata. Large extinct reptiles such as ichthyosaurs, dinosaurs and pterosaurs were only found in strata of Mesozoic age. However, when it came to matching contemporary strata from place to place, it became clear that only certain kinds of fossils were suitable.

For such correlation, organisms need preservable hard parts that commonly occur as fossils and are widespread in strata on a regional or, ideally, global scale. The fossil remains should also be readily identifiable and represent organisms that have a restricted time range – those that evolved rapidly. The first fossils that were recognized in the 1850s as satisfying most of these criteria were the Mesozoic ammonites described by the German palaeontologist Albert Oppel (1831–65). These free-swimming cephalopods are common and widespread in many

Albert Oppel was a German palaeontologist who first recognized that some fossils, such as ammonites, could be used to make very detailed and fine subdivisions within the succession of strata.

marine rocks. Their planispirally coiled shells are reasonably well preserved, can be relatively easily identified and, most important of all, evolved rapidly. Changes in ammonite species through successive Jurassic strata could be used both to finely subdivide the period and to match rocks of contemporary age across Europe. Subsequently, these fossils were used to make important correlations of Triassic strata throughout the geologically complex European Alps and in regions as far away as the Himalayas, Spitzbergen, eastern Siberia and California.

In Britain, a Scottish schoolmaster named Charles Lapworth (1842–1920) found that another extinct group of fossils, the graptolites, could also be used to subdivide and correlate strata, in this case across the structural complexities of the lower Palaeozoic outcrop in the Southern Uplands where he lived and worked. Lapworth went on to become professor of geology at Mason College, later the University of Birmingham. He pursued his graptolite studies with the aid of graduate students, especially two remarkable young women by the names of Miss Gertrude Elles and Miss Ethel Wood. Over a decade at the beginning of the twentieth century, they worked their way on foot, by bicycle and pony trap, through the Silurian and Ordovician strata of Wales and Cumbria, mapping the rocks and collecting graptolites. Because many of the strata looked similar, the only way the team could subdivide them was by identifying the graptolites that they contained. Following Oppel's example, Lapworth, Elles and Wood developed a graptolite biostratigraphy for the subdivision and correlation of Ordovician and Silurian marine strata. Although their scheme has now been substantially developed, it still forms the basis for global correlation of marine strata of this age. Since then, many other biostratigraphically useful fossil groups have been identified, especially microfossil groups such as the radiolarians, foraminiferans and extinct conodonts for marine strata, and plant pollen and spores for terrestrial and near-shore deposits. The advantages of microfossils are that they are often very abundant and can be recovered and identified from borehole samples. This is particularly important for hydrocarbon exploration.

The subdivision of terrestrial strata and their correlation with contemporary marine strata are often very difficult, as few biostratigraphically useful organisms are found in both environments. Terrestrial plant pollen and spores are important exceptions because they can be carried out to sea by wind and river water to be deposited alongside marine fossils. However, even pollen and spores may be very rare in some terrestrial deposits, often because they have been destroyed by oxidation in arid or semi-arid environments. Consequently, the biostratigraphy of some terrestrial strata, such as the Cenozoic strata of Eurasia, North America and Africa, makes use of various mammal teeth, especially those of rodents and pigs.

Fossils as evidence for evolution

One of the secretaries of the Geological Society of London between 1838 and 1841 was a young man by the name of Charles Darwin (1809–82). He had recently returned from a five-year round-the-world trip collecting all manner of natural history materials, including rocks, minerals and fossils, and he rather fancied himself as a geologist. Fellowship of the society had brought him into contact with the leading geologists and fossil detectives of the day. Darwin was in the lucky position of having something to 'sell' – his specimens. Not that he wanted money for them, but it was an important exchange of goods for recognition and status within the scientific community. He had already

In lower Palaeozoic rocks, fossils of extinct graptolites such as this Ordovician Didymograptus can be used to identify and match fine subdivisions of strata.

corresponded with several powerful and influential figures and was well advised about the process of handing out his specimens to various specialists for identification.

Darwin's unpaid job as secretary required his attendance at meetings, with the result that he became a familiar and well-known figure in the society. As he listened to presentations of the latest research and the debates that followed them, Darwin soon learned about the ins and outs of the fossil record and the attitudes and thoughts of the main players concerning its history. As Darwin began to mull over his gradually emerging ideas about evolution, he became increasingly aware that he was not going to get much in the way of support from the fossil record or from the fossil detectives of the day.

Fossil remains of giant Mesozoic reptiles were all the rage from the 1820s onwards as new finds were made and eminent fossil detectives, such as William Buckland (1784–1856) and ambitious rising stars like Gideon Mantell (1790–1852) and Richard Owen (1804–92), argued the toss about their true nature. By 1842, Owen had finessed them all by being the first to distinguish the Dinosauria as an extinct and distinct group of reptiles. Darwin saw just how ruthless and ambitious Owen could be and so was well aware that it would be unwise to encroach upon territory of which he was not a complete master – if he did, the very sharp and knowledgeable Owen would make mincemeat of him. Darwin was not a natural orator or debater. He preferred to construct his arguments in writing.

In 1844, the anonymous publication of a book called *Vestiges of the Natural History of Creation* caused an enormous public stir. The book was a hugely ambitious synthesis of information from across all the natural sciences in support of the idea of evolution. It argued that life could be created in the laboratory and that humans had evolved from apes, and presented a welter of information to support its thesis. It came just at the right time in Victorian intellectual development. There was a growing thirst for knowledge, especially about the natural world, with an explosion of cheap popular books about rocks, fossils, minerals and the history of life. Progressive ideas about evolution had been around since the end of the eighteenth century but were generally associated with Revolutionary France and therefore considered inherently anti-religious and a danger to the stability of society.

Part of the huge success of *Vestiges*, which sold 40,000 copies in Britain, was a result of the gossip that developed around its authorship, which was a remarkably well-kept secret. Not until 40 years later, in 1884, was it revealed that the author was an Edinburgh publisher by the name of Robert Chambers (1802–71), and by then he had been dead for 13 years. Scientifically informed opinion was very divided about the book, with many

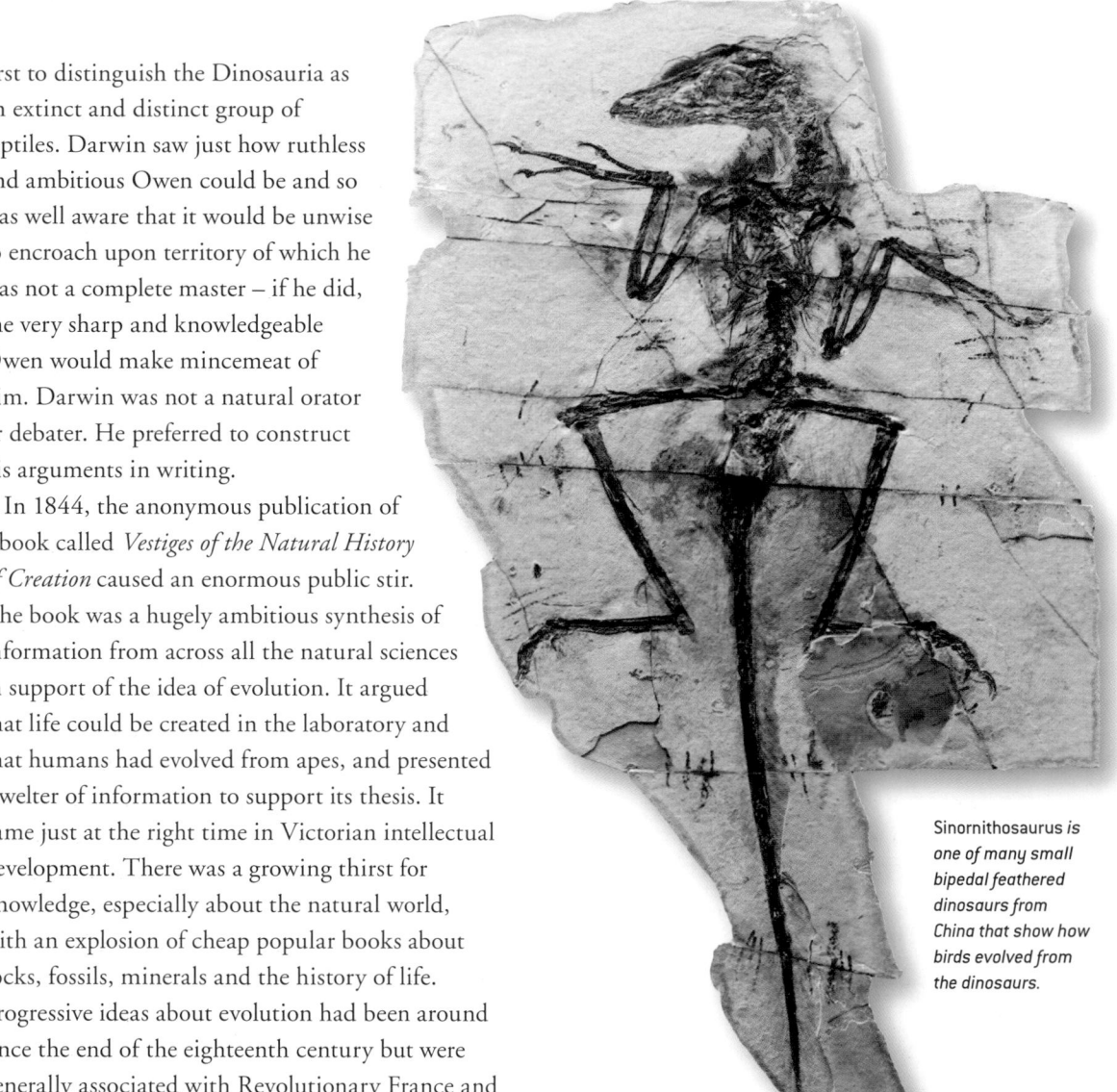

Sinornithosaurus *is one of many small bipedal feathered dinosaurs from China that show how birds evolved from the dinosaurs.*

vociferous and severe critics who attacked its speculative and often inaccurate science as well as its anti-Creationist stance. Darwin was made aware of the strength of the criticism, who it was coming from, and of the nature of the argument against evolution. Being so forewarned, he was better equipped to head off similar criticism for his own developing theory of evolution. He realized that one of the subjects he had to be particularly careful about was the fossil and rock record. Many of the eminent palaeontologists of the day were strong opponents of the idea of progression and particularly opposed to any suggestion of human evolution.

Despite all Darwin's precautions, when *On the Origin of Species by Means of Natural Selection, or The Preservation of Favoured Races in the Struggle for Life* was first published in 1859, it did indeed attract the criticism of the usual suspects such as Richard Owen. Owen tried to use the evidence of his newly 'invented' dinosaurs as part of his attack against evolution. He argued that since the dinosaurs were clearly a very successful group of large animals, apparently well adapted and 'fit' in evolutionary terms, they should, according to the theory of evolution, have progressed and evolved; instead, they died out. The main problem that the fossil record seemed to present to Darwin was the lack of evidence for organisms that were 'intermediates', or more correctly common ancestors, between major living groups. According to the theory outlined in *On the Origin of Species*, there is an expectation that life has diversified and descended from common ancestors, so where were the fossils that represented the common ancestors of the birds and reptiles or of the mammals and reptiles, for example?

Fortunately for Darwin, it was only a year after the publication of *On the Origin of Species* that a very important piece of fossil evidence turned up in a quarry in southern Germany. At Solnhofen the late Jurassic limestones were worked as lithographic stones, and are still worked for paving and ornamental slabs today. An extraordinary range of fossils has been recovered from these fine-grained lagoon deposits, and the quarrymen have traditionally put aside good specimens for sale. So it was with the small 5-cm-long feather, beautifully preserved in monochromatic carbon, and very clearly a modern-looking asymmetrical flight feather. The quarrymen

knew that, where there is a feather, there should be a bird, and a year later they found it.

It was Richard Owen who heard of the find and used his considerable influence to persuade the British Museum to buy it. In 1862, *Archaeopteryx* (meaning 'ancient wing'), as it had been named, was purchased along with 1700 other Solnhofen specimens for the not inconsiderable sum of £700 from Dr C.F. Haberlein, a local amateur fossilologist. He used the money to provide a dowry for each of his six daughters. Once in possession of the fossil, Owen produced a masterly description of its anatomy, showing that it has a curious mixture of reptilian features, such as a long bony tail and toothed jaws, along with avian features, such as feathers and wings.

It was Darwin's young supporter Thomas Henry Huxley (1825–95) who seized upon this evidence as support for evolution. He claimed that this extinct reptile-bird demonstrated that 'missing links' were no longer entirely missing and that the fossil record could be expected to provide more in the future – as indeed it has done. Darwin's theory was the major turning point that led to our modern understanding of the evolution of life on Earth. We now know, thanks to some remarkable fossil finds in China for example, that feathers are not unique to birds but first evolved in the dinosaurs. Indeed, we can now see that dinosaurs are not extinct – they are covered in feathers, and we call them birds!

Darwin would no doubt be enormously gratified if he knew the extent to which fossil discoveries over the intervening years have supported the theory of evolution. Now we have very detailed fossil evidence for speciation in many different kinds of organisms, as well as wider-scale evidence for the evolution and extinction of major groups over time. Even at the biomolecular level, the recovery of ancient DNA from fossils is providing evidence of relatedness in a variety of organisms, including between the Neanderthals and modern humans.

Forensic palaeontology

To get as much information and satisfaction as possible from fossil-hunting, it is important to go about the business with as much care and attention as a detective might use to get a good result from an investigation.

In the historic past, fossil-collectors were only interested in removing a good specimen from the rock, cleaning off any residual material and placing the prize in a glass case. At best, they might have added a label saying where the fossil came from or some vague indication of its age. Today, we know that fossils can provide us with so much more valuable and interesting information and still be fascinating objects in their own right.

Is the fossil original or a 'plant'?

Like a crime scene, the collecting site should be carefully assessed for any possible contamination that would be misleading. Is the potential fossil material in its original place or has it been brought into the site? A regular student prank on field trips is to collect a fossil from one place and then plant it at another site to try to fool the leader, other students or an unwary collector. This might seem like a bit of fun, but historically, planting of information of this kind has been much more damaging and sometimes even malicious, as in the infamous Piltdown find of 1912 (see p. 126).

Once the fossil detective has taken careful note of any possible false leads originating from human intervention, the site can be assessed for other potential problems. Most fossil evidence is still buried in its rock matrix. Rather than immediately searching the rock for fossils, the fossil detective should carefully examine the rock itself to see what clues it can provide about the original depositional environment.

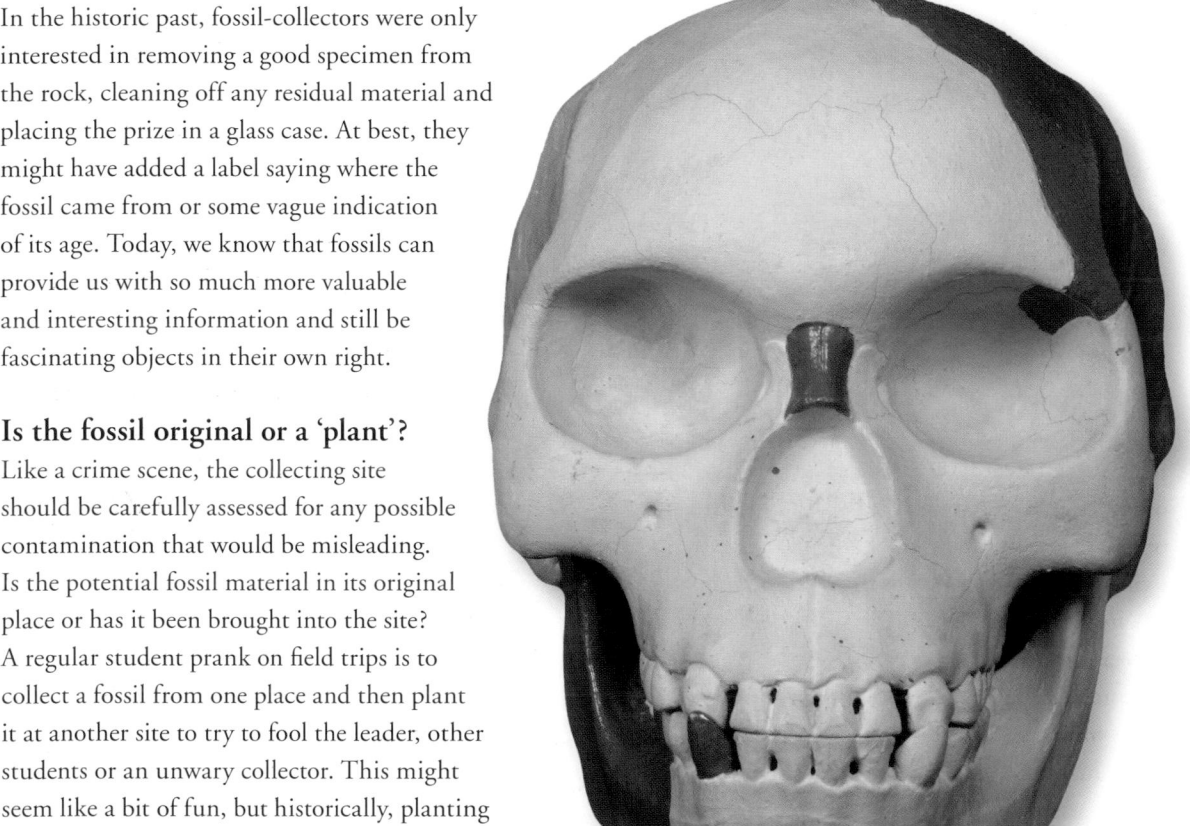

Despite very few fossils at the Piltdown site, a skull was reconstructed for Piltdown Man from the bits of bone (brown coloured) found there.

What was the original context of the burial?

There are many questions that need to be answered. For instance, what kind of sedimentary rock is the fossil found in? What is the succession of strata at the site? Are there other layers that might provide important clues concerning the origin and age of the

strata? For instance, there may be some sandstones with well-developed internal structures, known as cross-bedding, that will immediately suggest whether the original environment was on dry land or under water. River and delta deposits have characteristics that can be identified once the signs are known. While this information is available in books, it is much better acquired through experience in the field with expert guidance.

Was the body moved?

Only once you have a reasonable idea of the geological setting of the particular strata you are interested in should you proceed with your investigation. And even then, there are more questions to be answered. Did the sediment originate locally or was it transported from far away? This is important because the fossil evidence may also have been transported. Sometimes rocks contain mixed and potentially confusing fossil evidence, some of which has been transported, and some of which is more locally derived. Evidence of physical transport by currents (such as wind or water) can usually be spotted quite easily because the processes tend to dislocate, break and abrade fossil material.

Has the fossil's appearance been altered?

Once you have some idea of the origin of the fossil evidence, there is yet another layer of unravelling to do. What is the state of preservation of the fossil material? Does it have its original shape and composition, or has it been altered – and, if so, in what way? Post-mortem processes of burial and fossilization, along with other geological events such as folding and metamorphism, can radically alter the appearance of fossil remains. In extreme circumstances, they can make it impossible to identify the remains, although there are now techniques that help to recover valuable information that might otherwise be lost for ever.

Replicating the body

There are now various computer-based techniques for restoring the original shape of fossils that have been distorted by rock movement. Where the original shell of skeletal material has been dissolved by natural processes, the form can also be replicated remarkably faithfully by making a cast from the natural sediment mould.

There are a range of casting materials, from plasticine to various silicone, plastic and rubber

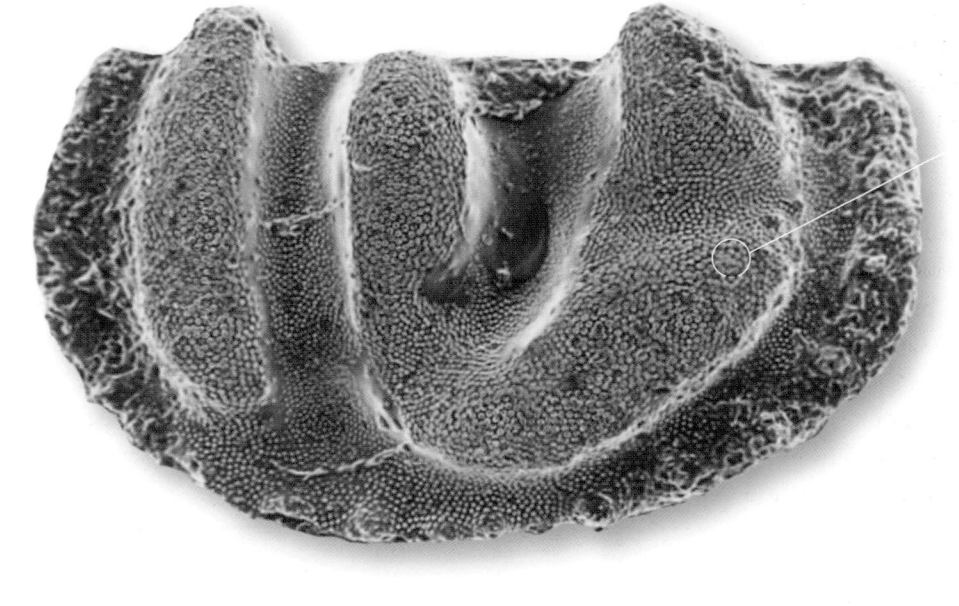

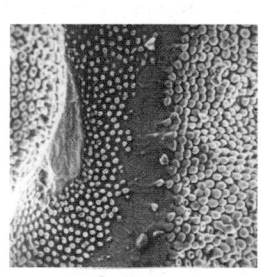

Scanning electron microscopy can reveal enough details of millimetre-sized fossils such as this Silurian ostracode for the species to be identified accurately.

materials whose use depends upon the shape of the fossil and nature of the mould material. Fine-grained sediment moulds, such as those formed by clay, can replicate surface detail down to micron-scale cellular structure. The advantage of making such casts is that they can be replicated and examined in electron microscopes and it does not matter if they are damaged in the process because, unlike an original specimen, they can be replaced easily.

Altered states

The process of fossilization often radically alters the chemical composition of the original shell or skeletal material of the fossil. Historically this was one of the main reasons why it took so long for fossils to be recognized as the remains of once-living organisms. There are many different chemical changes that can radically alter the appearance and composition of fossil remains. Original bone, shell and woody materials can be replaced by a variety of inorganic materials from siliceous opal to metal minerals such as pyrites (iron sulphide, also known as 'fool's gold') and crystalline minerals such as quartz. In the process the original internal structure is often destroyed.

For all of these reasons, the preparation and assessment of the locus of your find can be a painstaking and lengthy process before you even get to grips with the essential fossil evidence. In the meantime, however, you will have already come a long way in understanding how and why the fossil reached its location and why it has its present appearance and state of preservation, which is often far removed from the original state.

Position of the body

The position that any fossil has within its enclosing sediment provides clues as to whether it is found in its original location, or whether (and how) the remains have been moved. In order to answer these questions it is necessary to have a general knowledge of the life habits of a range of fossil animals and plants. Often the form of the body itself provides good clues as to the organism's life habits. For example, the shell form of common marine bivalves shows whether they lived freely on the sediment surface, attached to some object, or burrowed into the sediment. Burrowers tend to have streamlined shells (for example, the living razor shell), while free-living forms on soft sediments tend to have wide and flat saucer shapes (for example, the living scallops).

A number of tests can be applied to check whether the remains are found in their original location. For example, are the two valves of bivalved forms such as clams and brachiopods still articulated, in which case they may be close to home? Or have they been broken apart, in which case they have probably been moved? Are shells of different size present, in which case they may be in place? Or are they mostly of one size, in which case they have probably been sorted by current

This ammonite's original shell material has been replaced by 'fool's gold' (iron pyrites), a mineral that commonly forms in muddy sea-bed sediments.

Fossilization processes have preserved the internal sediment moulds of these bivalve shells, which are revealed as the original white shell material has been worn away.

activity? Current activity also tends to orient any elongate shells as well as move them.

Are there any suspicious marks?

The evolutionary arms race between predator and prey goes back a long way. Even one of the first fossil shelled organisms, the calcareous tube worm *Cloudina* (from latest Neoproterozoic times, some 547 million years ago), has a neat hole bored in its shell by some unknown predator. It is clear that the calcareous shell was not enough protection from predation. Ever since, perforations such as this are commonly found in shells. They are produced by a variety of predators, but known to be especially caused by gastropods.

In more modern times, powerfully clawed crustaceans can break thin shells with distinctive fracture patterns. However, you need to find the pieces to be able to identify them and also to differentiate between a claw fracture and other natural breakage patterns, which can be tricky. Many vertebrate animals – ranging from birds (such as oyster catchers) to otters, as well as several extinct groups of marine reptiles – are specialist feeders on molluscs. Some of these predators make their mark with identifiable damage patterns.

Plants have also been involved in the arms race and have evolved all sorts of strategies to overcome lethal damage by predators, especially the production of an amazing array of toxins that help protect their leaves. But evidence of attack, especially by insects, is common in fossil leaves, which have chew marks, galls and the burrows of leaf miners.

Evidence of tooth and claw marks

Tooth marks are often found on the bones of land-living vertebrates, but the problem in such cases is differentiating between the marks of the 'killer' or the marks of the opportunist scavenger. Even the infamous *Tyrannosaurus rex* is often portrayed as more of a scavenger than an active predator. However, recent biomechanical studies of the skull and jaw musculature of this giant dinosaur show that the force of its bite was so great that it was more likely to have been an active predator. Again the context of the bones and the position of the bite or chew marks can determine whether these marks occurred in a fight or were made post-mortem by scavengers.

LONDON

London might seem a long way from the traditional fossil-hunting grounds of the North Yorkshire coast or the cliffs at Lyme Regis, but the city's centre is full of fossils, if you know where to look. Fossils can be found in many of the rocks used to build the city — which came from all over Britain and abroad. What at first might seem like unidentifiable squiggles and swirls in polished façades are often shells, sponges, crinoids or corals dating back millions of years. Once your eye becomes accustomed to the shapes, it is not hard to identify the distinctive Jurassic gastropod known as the 'Portland screw' in Portland Roach from Dorset, or the circular outlines of a rudist bivalve colony in Cretaceous limestone from Europe. Guided geology walks are a great way to find out more about the variety of fossils in this urban environment.

For fossil fanatics, no trip to London would be complete without a visit to the Natural History Museum in South Kensington. Few can fail to be impressed by the stunning cast of a 26-metre-long *Diplodocus* skeleton that greets you as you enter. The museum opened in 1881 and houses one of the most outstanding palaeontological collections in the

A view of the River Thames shows the Houses of Parliament on the north bank and the London Eye on the south. Museums in London, particularly the Natural History Museum, have built up world-class collections of fossils, including an amazing variety from Britain.

The Crystal Palace Dinosaurs

THE WORLD'S FIRST THEME PARK

After the Great Exhibition in 1851, plans were made to relocate the famous Crystal Palace from its original location in Hyde Park to Sydenham in the south of the city. Having invented the term 'dinosaur' just ten years earlier, Professor Richard Owen (1804–92) saw an opportunity to create a visitor attraction of life-sized prehistoric animals as part of the new Crystal Palace Park. Owen enlisted the help of the artist and sculptor Benjamin Waterhouse Hawkins to create his ambitious scheme for a display of more than 15 different types of extinct animal, including giant amphibians, crocodiles, marine reptiles, pterosaurs, early mammals and dinosaurs. Giant models of the animals were to be arranged around a series of artificial lakes and landscaped islands grouped with appropriate rocks and plants according to their geological age.

world. In the Fossil Marine Reptiles and Dinosaurs galleries you can see some of the best specimens of these giant extinct animals ever found in Britain. The museum was the brainchild of Richard Owen, the nineteenth-century anatomist who identified the fossilized remains of many animals and coined the term 'dinosaur' in 1841. By many accounts, Owen was an unlikeable character, but the museum is a credit to his ambitious nature.

Hawkins worked closely with Owen to get the proportions and shape of his models right. Among them were the first three dinosaurs known to science: *Iguanodon*, *Megalosaurus* and *Hylaeosaurus*. It was still early days in the study of dinosaurs. Very little was known about these creatures and there was no consensus as to what they looked like. This was the first ever attempt to reconstruct life-sized models of them. The final designs were based on Owen's interpretation of the limited fossil evidence that was available to him. As no complete skeletons of any dinosaur had yet been discovered, Owen put his own stamp on their appearance. He correctly interpreted the dinosaur pelvis and put the legs under the body

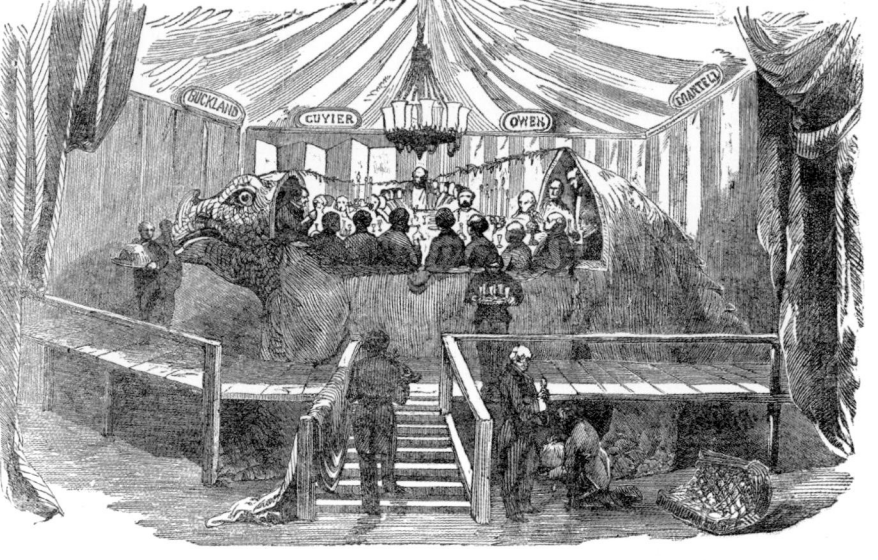

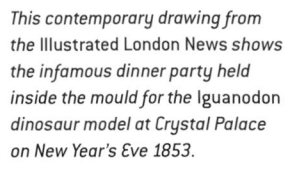

This contemporary drawing from the Illustrated London News *shows the infamous dinner party held inside the mould for the* Iguanodon *dinosaur model at Crystal Palace on New Year's Eve 1853.*

rather than giving them a sprawling, lizard-like stance. However, the idea that such an enormous beast could walk semi-upright on two legs with its tail held aloft was still years away. The Crystal Palace dinosaurs stand firmly on four solid, column-like legs, and the *Iguanodon* looks more like a rhinoceros than modern reconstructions of the actual dinosaur.

Hawkins set up a temporary studio on the construction site in 1852. His enormous sculptures attracted a huge amount of interest well before they were finished. Both he and Owen knew the value of publicity if their venture was going to be a success; one of their most famous stunts was to host a dinner party for 20 distinguished guests inside the mould of the *Iguanodon*. The party took place on New Year's Eve 1853 and included a seven-course, fossil-themed menu and a special song composed for the occasion. Apparently, there were loud renditions of the chorus sung late into the night:

'The jolly old beast
Is not deceased
There's life in him again!'

The invitation card, designed by Hawkins, was a work of art in itself, with the words written on a drawing of an outstretched pterodactyl's wing. The party was a resounding success and was featured by the *Illustrated London News* and in *Punch* magazine under the title 'Fun in a Fossil'. The official opening of the Park by Queen Victoria in 1854 drew a crowd of 40,000 visitors, with the prehistoric animal trail, the world's first theme park, as the star attraction. When the dinosaurs were unveiled they caused quite a sensation; scientists were impressed, adults were awe-inspired and children were terrified! Nothing quite like them had ever been seen before.

By today's palaeontological standards, Owen's models look very strange indeed. The plesiosaur models, for example, have impossibly twisted necks, and the ichthyosaurs are wrongly shown with their large sclerotic eye sockets exposed, no dorsal fin and a flat tail. Both types of creature were positioned half out of the water, basking on land like seals. Admittedly, they would have been hard to see had they been fully submerged, but we now know that they were like modern-day dolphins and never came ashore. One of the oldest animals depicted was *Dicynodon*. The two *Dicynodon* models correctly have two long teeth on the upper jaw, but Owen and Hawkins gave them shells and a turtle-like stance. They only had a fossil skull and a handful of bones to go on, and the shell was an educated guess, but no *Dicynodon* discoveries

since have ever revealed a shell. The most accurate models were of recently extinct animals, such as the Giant Elk, *Megaloceras*, and the giant ground sloth, *Megatherium*.

Mini versions of the models and posters of the animals were successfully marketed at home and abroad. In 1866, Hawkins was asked to create a similar display in New York's Central Park. This venture would have included two new recent dinosaur discoveries made in the USA, *Hadrosaurus* and *Laelaps* (known today as *Dryotosaurus*). After setting up a studio and working on several models, Hawkins' plans fell victim to a change in local administration and corrupt politicians. The models were smashed up and are rumoured to be buried under Central Park to this day, but have never been found.

Hawkins built his statues to last, and after more than 150 years they are still standing in their corner of Crystal Palace Park, exactly where he and Owen positioned them. They are a wonderful testimony to the vision and ambition of Victorian science. After a major renovation project in 2003, the prehistoric animal trail once again attracts thousands of visitors a year. The painted statues of dinosaurs and other prehistoric creatures, with their concrete scaly skin, may be rather out of date, but that is part of their charm. They are a reminder that palaeontological ideas have changed over the years, in the same way that the fossil animals under scrutiny have evolved through time. The models may be inaccurate, but they provide a marvellous insight into why dinosaurs initially captured the public's imagination, and have continued to do so ever since.

Many of the original models can still be seen in Crystal Palace Park today, including the dinosaurs Iguanodon (top) and Megalosaurus. Both models stand on four sturdy, column-like legs. We now know that these dinosaurs had a much more athletic posture and balanced on their hind legs, which were significantly larger than their forelimbs.

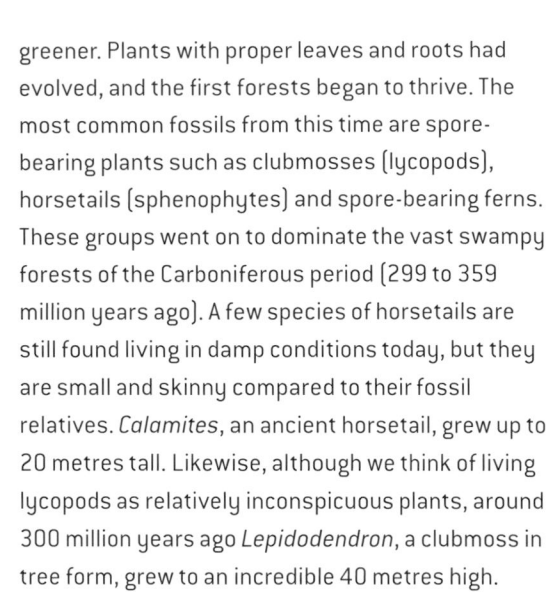

The Evolution of Plants

A POTTED HISTORY

Just like animals, plants evolved from very small and humble beginnings billions of years ago into the staggering variety of forms alive today. The vast majority of living plant species are angiosperms (flowering plants), but the fossil record reveals that these are simply the most recent in a series of plant groups that have dominated Earth at different times. Plants do not fossilize as well as animals because they have a softer structure and lack hard parts like bones and teeth. They often break up when they die, which means that the fossil leaves, spores, seeds, roots, stems and branches of a single plant species are hard to match, and often have different names. Nevertheless, the huge diversity of rocks in Britain means that the evolution of plants can be effectively traced through time using fossil evidence.

Plants originated from green algae, and made their first transition from life in water to life on land around 450 million years ago. This new environment posed some interesting problems for early plants, such as how to avoid drying out in the sun and how to grow upright without any support. *Cooksonia* was one of the original land plants to adapt to these challenges. It is the first well-known fossilized vascular plant, which means it had specialized cells to transfer water and nutrients through its structure as most plants do now. It belonged to a group of plants called the rhyniophytes that are now all extinct.

By Devonian times (359 to 416 million years ago), the barren, rocky landscapes of Earth were noticeably greener. Plants with proper leaves and roots had evolved, and the first forests began to thrive. The most common fossils from this time are spore-bearing plants such as clubmosses (lycopods), horsetails (sphenophytes) and spore-bearing ferns. These groups went on to dominate the vast swampy forests of the Carboniferous period (299 to 359 million years ago). A few species of horsetails are still found living in damp conditions today, but they are small and skinny compared to their fossil relatives. *Calamites*, an ancient horsetail, grew up to 20 metres tall. Likewise, although we think of living lycopods as relatively inconspicuous plants, around 300 million years ago *Lepidodendron*, a clubmoss in tree form, grew to an incredible 40 metres high.

The major coal deposits in Britain that fuelled the Industrial Revolution in the eighteenth and nineteenth centuries were formed from the compressed and fossilized remains of the Carboniferous forests. Fossil leaves reveal that oxygen levels were higher in the Carboniferous period than at any other time in Earth's history. Palaeobotanists work this out by examining the density of the pores on fossil leaf surfaces, which control how plants exchange gases with the atmosphere. Today, oxygen makes up about 20 per cent of the atmosphere, but back then it was more like 30 per cent. Giant dragonflies, with a wing-span of up to a metre, evolved in the oxygen-rich atmosphere. It is thought they would not be able to fly today, due to our relatively oxygen-poor atmosphere.

During the early Mesozoic era that began 251 million years ago, another category of plants, called the gymnosperms (meaning 'naked seed'), took over as the dominant flora. This group forms the majority of plant fossils in rocks of this age. Cycads, conifers, seed-ferns and bennettitales (an extinct order of seed plants) flourished, and descendants of all these types, except the bennettitales, make up a small

Orchids growing in the Prince of Wales Conservatory at the Royal Botanic Gardens, Kew. The fossil record shows that flowering plants (angiosperms) began their rise to dominance about 100 million years ago in the Cretaceous period, and by Tertiary times all the major groups had evolved.

Cooksonia was one of the earliest vascular land plants to make the transition from living in water to living on land. It had bifurcating stems topped with spore sacs, seen here on Cooksonia petroni, 10 mm in length, from early Devonian rocks in Herefordshire.

The fossilized remains of a Mariopteris fern. This specimen is from the Carboniferous Coal Measures in the Nottinghamshire coalfield, and is about 14 cm long.

proportion of today's flora. In the mid- to late-Cretaceous period (approximately 100 million years ago) the angiosperms, so familiar to us today, began to take over. By the beginning of the Tertiary period 65 million years ago, all the major types of flowering plants, including the precursors to the grasses, had evolved. The world's landscapes have been covered by this familiar flora ever since.

There are currently thought to be over 250,000 living species of angiosperms, compared to 10,000 species of pteridophytes (ferns) and just 750 species of gymnosperms, including all conifers. The changing fortunes of different plant groups throughout earlier periods of Earth's history have been largely overshadowed by the runaway success of the angiosperms. What interests palaeobotanists is why the angiosperms appear quite abruptly in the fossil record and were seemingly able to out-compete all other types of plants so efficiently, despite evolving relatively late in the history of plants. Co-evolution with the insects needed to pollinate them seems to have been the key to their early success, and since then co-evolution with birds and mammals has enhanced their grip on the planet even more. Virtually all the plants we now cultivate for food are angiosperms.

In London, there is no better place to appreciate the variety of living plants than the Royal Botanic Gardens at Kew, near Richmond. You will not find many conventional fossils at Kew but in the Evolution House, laid out under one roof, the development of plants can be traced as far as is possible via living specimens. Do not be fooled by the *Cooksonia* or *Lepidodendron* on display, though: they are plastic models! Many of the living plant types on display can be described as 'living fossils', which means that there are fossilized examples of the same plant species, or a very closely related species, dating back many millions of years.

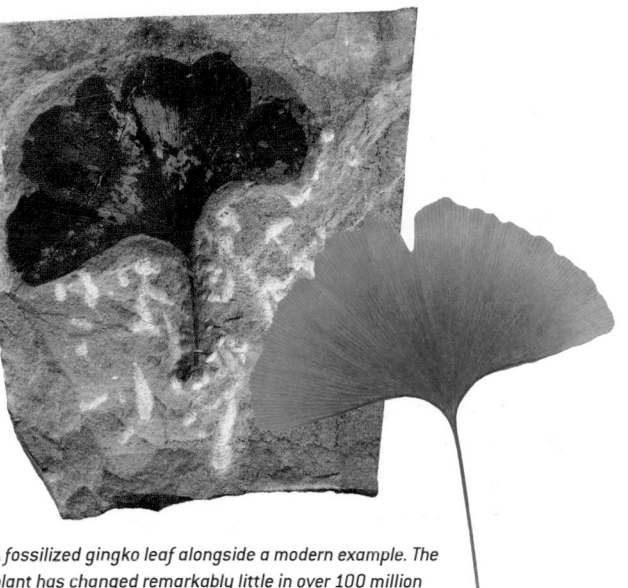

A fossilized gingko leaf alongside a modern example. The plant has changed remarkably little in over 100 million years and is an example of a living fossil.

A classic example of a living fossil is the ornamental tree *Gingko biloba*, more commonly known as the Maidenhair tree. It is very rare in the wild but a common tree in parks and gardens. It is the only surviving species of an entire group of plants known as the gingkophytes that flourished during the Mesozoic period. Very distinctive fossilized leaves suggest that *Ginkgo biloba* has hardly changed for over 100 million years. Another intriguing living fossil on display at Kew is the Wollemi pine (*Wollemia nobilis*), which can be traced back around 90 million years via similar-looking fossils. It was thought to be extinct until 1994, when about 100 living specimens were discovered in a deep, almost impenetrable gorge near Sydney in Australia. Several specimens were planted at Kew as part of a propagation and conservation programme, and they appear to be thriving. Living fossils may not be fossils in the conventional sense but they are still a fascinating way to explore ancient life. They are also a reminder that, like animals, plants evolved into a wide variety of forms as they adapted to the different environments found on Earth.

The presence in the London area of fossils of the same species of hippo – Hippopotamus amphibius – as live in parts of Africa today is evidence that there were dramatic shifts in Britain's climate during the Ice Age.

Hippos at Home

LIONS IN TRAFALGAR SQUARE

These days the only lions you are likely to see in Trafalgar Square are the magnificent bronze statues that commemorate Nelson's victory in the eponymous battle of 1805. But if you had been walking in the area 125,000 years ago, you might have encountered a real one. Fossil bones and teeth dug up during building work in the 1950s revealed that animals such as lions, hyenas, elephants and rhinos, which we now think of as exotic, once populated this region in large numbers. Fossil fragments of the hippopotamus stood out as being especially abundant, and we now know that it was the most common big animal in Britain at the time.

The hippos that swam in the River Thames thousands of years ago were the same species, *Hippopotamus amphibius*, that lives in Africa today. Just as they do now, herds of hippos would have basked in the river by day and grazed on the grassy banks at night. Their grazing habit would have kept a wide strip of land either side of the river free of trees, making it a popular feeding spot for animals such as deer, and a prime hunting spot for lions and hyenas. Fossils of the same age from North Yorkshire and Devon show that hippos were quite at home all over Britain, not just in London.

Hippos, lions and hyenas lived on the banks of the Thames 125,000 years ago. The area around Trafalgar Square would have looked something like this.

The London hippo fossils were found in gravels laid down by the River Thames during the Ice Age. Despite its name, the Ice Age was not always icy: the climate switched between cold phases known as 'glacials', when ice sheets covered much of Britain, and shorter, warmer phases known as 'interglacials'. The size and position of the Thames has varied during the past 2 million years, and the river has deposited a thick sequence of gravels across London and parts of Essex and Kent. These gravels contain fossils that reflect the changing environmental conditions of the Ice Age. The gravels under Trafalgar Square were deposited during an interglacial called the Ipswichian that lasted for approximately 60,000 years and peaked around 125,000 years ago. Plant fossils found alongside the animal remains showed that species such as maples (*Acer monospessulanum*) and water chestnut (*Trapa natans*), which are more at home in southern Europe today, grew in London at that time. Temperatures were two to three degrees Celsius warmer compared to today. We know that hippos were well suited to this climate (although it was slightly cooler than Africa is now), but how did they get here?

As the climate of the Ice Age fluctuated, so did the geography of the region. During cold phases, sea levels were much lower and Britain was connected to mainland Europe by areas of dry land. Animals and plants were able to spread north or south via these land bridges to keep up with the changing environmental conditions. As temperatures warmed at the start of the Ipswichian period, hippos must have migrated north to Britain when sea levels were still low enough to cross from mainland Europe. Once sea levels rose and Britain became an island, the hippos continued to live here until the next cold phase. To date, scientists have been unable to work out the route the hippos took, but they were certainly here in large numbers. Interestingly, hippos seem to have been rare in Britain during other interglacials.

River gravels laid down by the Thames at the same time as those found beneath Trafalgar Square are exposed along the foreshore at East Mersea in Essex. Occasionally, fossil fragments of hippos and other Ice Age mammals can be found on the beach.

Since the famous discoveries beneath Trafalgar Square in the 1950s, palaeontologists have studied fossils in the river gravels at many different sites across the region. They have been able to piece together a detailed picture of the flora and fauna of Ice Age Britain, not just during the Ipswichian period, but during other warm and cold phases, too. What they have found is that each interglacial in Britain had a distinct flora and fauna. In Ipswichian times, for example, hippos were dominant, whereas horses were completely absent; during other interglacials the reverse was true.

When conditions changed at the beginning of each interglacial, it seems that some animals managed to migrate north fast enough before sea levels rose, while others didn't make it. Varying rates of change have dictated which animals lived in Britain at different times.

Opportunities to investigate the Ice Age flora and fauna of central London today are limited to the few major building projects that dig down into the Thames gravels. But outside the city, along the Essex coastline, for example, there are places where you can hunt for Ice Age fossils. The so-called hippo gravels, like those investigated under Trafalgar Square, are exposed along the foreshore at East Mersea. You might be lucky enough to find a hippo bone fragment washed up on the beach. The best of the British Ice Age fossils are on display at the Natural History Museum — yet another reason to visit next time you are in London. And when you cross over the River Thames, imagine hippos wallowing in the mud just as they did not all that long ago.

The Tertiaries
OF THE LONDON BASIN

The city of London lies in a broad structural basin that also carries the River Thames. Broadly speaking, the youngest surface deposits of the post-glacial, Holocene age lie in the middle of the valley. These are surrounded by older strata from Quaternary Ice Age deposits down into those of the Tertiary (Cenozoic) age. Beneath these lies the Cretaceous chalk that forms the rolling hills of the Chalk Downs on both the northern and the southern sides of the basin and valley.

The Tertiary deposits and their fossils tell an interesting story. The oldest sediments are shallow sea sands of the late Paleocene age (58 to 55 million years ago), laid down in warm, seasonally dry climates as the sea flooded over the late Cretaceous chalk landscapes. Their deposits can be seen along the Thames estuary foreshore – at Herne Bay, and at Pegwell Bay on the Kent coast, for example. Analysis of the sand shows that the sediment was brought south from the Scottish Highlands by rivers draining into the North Sea basin. At this time there were large-scale tectonic movements as the North Atlantic opened, with volcanoes erupting in northwest Scotland. Southern England was also affected by the nothernmost ripples of the folding and faulting of the European Alps.

Fossilized seashell and fish remains from these Paleocene strata, and deposits from the slightly younger Eocene age (55 to 34 million years ago), have been known for many years. The fish were mostly small sharks, such as sand sharks (*Synodontaspis*), plus dogfish (*Squalus*) and rays. Over the years fossil detectives have found about 20 different species of shark in the London Basin. As sharks and rays are cartilaginous rather than bony fish, most of their remains are teeth. Although one shark's tooth might at first look much like another, different species can be identified by their teeth alone.

Abbey Wood, in Greater London, is one of the most important sites of early Eocene age in Britain and consequently has legal protection as an SSSI.

For well over 150 years, the cliffs and foreshores between Herne Bay and Reculver Point have been famous for their fossiliferous sediments from the early Tertiary age.

A tooth from one of some 20 species of fossil shark known from the Isle of Sheppey.

Excavations in the pebbly sands here have revealed a remarkable fauna of shellfish, fish (including seven or eight shark species), reptiles (such as the soft turtle *Trionyx*), birds and 28 species of early mammals. Most of the vertebrate fossils are just teeth that have been sieved from considerable volumes of sediment. The mammals are remarkably diverse and include the small primitive browsing horse, *Pliolophus*, early primates such as *Cantius*, extinct creodonts such as *Oxyaena* and a bat, *Eppsinycteris*. Overall, the fauna represents a mixture of marine, brackish-water- and land-living creatures.

Evidently, this was an estuary in early Eocene times in which marine and brackish-water shellfish, fish and aquatic reptiles, such as turtles and crocodiles, lived. The nearby shoreline was a dense subtropical forest whose trees and shrubs were browsed by small primitive horses, while the canopy above was inhabited by small, fruit-eating primates.

Best-known of all the Eocene strata is the London Clay, which has been used for brick making since Roman times. When the Romans set up camp in London between 43 and 50 AD they used local timber because of the lack of local building stone. In an uprising against Roman rule, led by Queen Boudicca, the Iceni torched this first Londinium fort. Traces of the fire can still be found in London's subsoil. The Romans returned to avenge their humiliation and built a more substantial fort surrounded by a 3-mile wall of brick and stone, bits of which still remain. The Romans were expert at practical geology and knew which stones and clay would make bricks and cement. As clay was much more abundant locally than stone, the manufacture of the ubiquitous London brick dates from this time.

The proximity of such fossil riches near to London meant that there was a thriving trade in fossils to meet the demands of the metropolitan gentlemen collectors. By the nineteenth century, the fossil plants of the London Clay were well known, and since that time some 300 plant species, as well as fish, crocodiles, snakes and birds (among others) have been recorded.

Analysis of the changing sediments and faunas from Paleocene into Eocene times shows that around 55.8 million years ago, global temperatures rose by up to 10°C (50°F) over 10,000–20,000 years before returning to warm background temperatures. Such 'fossil' records provide us with good examples of what we can expect as a result of global warming.

Pyritised internal casts of fossil fruits and seeds from the Isle of Sheppey.

SCOTLAND

Scotland is home to the most dramatic scenery and the most diverse rocks of all Britain's regions. In fact, it is often said that no other country of its small size has such varied geology. Other Scottish claims to fame include some of the oldest rocks in Europe, as well as the oldest fossil evidence found anywhere in Britain.

Up in the far northwest of the country, rocks called the Lewisian Gneiss are a staggering 3.3 billion years old. Lying on top of these ancient metamorphic rocks are layers of sedimentary rocks called the Torridonian. They are much younger but some are still an astounding 1 to 1.2 billion years old. Within some of the sandstones, such as those found along the north shores of Loch Torridon near Diabaig, tiny wrinkles indicate that microbes lived on the sandy sediments that formed the rocks. These microbes were among the earliest pioneers of life on land. Over a billion years ago, they were in the vanguard of the incredibly diverse succession of Earth's animals and plants.

Probably the most famous Scottish story about prehistoric life is the myth of the Loch Ness Monster. Whether you believe in Nessie or not, the loch does contain surprising fossil evidence.

James Hutton

THE FOUNDER OF MODERN GEOLOGY

The fact that we can appreciate the evolution of life over thousands, millions and even billions of years is largely thanks to an exceptional Scotsman called James Hutton (1726–97). In the late eighteenth century, Hutton was the first person to truly grasp the concept of an immensely old Earth and prove that our planet had to be much, much older than previously thought. Current estimates put the age of Earth at 4.5 billion years: this is older than Hutton proposed, but not beyond the scope of his theories. He based his ideas on more than 30 years of geological observations and is widely revered as the founder of modern geology.

Hutton was born in Edinburgh in 1726, at a time when it was generally accepted that Earth was just 6000 years old. A precise age had been calculated by Archbishop James Ussher in 1658 based on biblical texts: Earth was created on the night of 23 October 4004 BC. In the eighteenth century, all knowledge about Earth was confined within a rigid religious framework. Challenging these long-held views was certainly not encouraged. A determined and very bright individual was required to initiate a fundamental shift in beliefs.

Loch Ness extends for 22 miles south of Inverness. The surrounding landscape was shaped by glaciers and ice sheets that melted away 10,000 years ago. Fossil evidence in the loch sediments records both natural and man-made events.

The loch is very deep – up to 230 metres – and within the lake-bed sediments, microfossils such as diatoms and pollen grains can be found that record 10,000 years of regional environmental history. Events such as the beginning of farming, industrialization, volcanic eruptions in Iceland and the 1986 Chernobyl disaster can be identified in cores taken from the lake bed. Sadly, there is nothing to suggest a monster ever lived there.

Hutton studied medicine at Edinburgh University and in Europe before following in the footsteps of his father to run a farm in

Among his many skills, Hutton was an accomplished geologist, naturalist, chemist and experimental farmer.

A view from Blackford Hill in Edinburgh looks towards Arthur's Seat (right), the eroded remains of an ancient volcano, and Salisbury Crags (left). These features dominate the Edinburgh skyline and were an important source of inspiration for Hutton.

Berwickshire. It was there that he developed a keen interest in soils and rocks. Despite the difficulties and hazards of eighteenth-century transport, he travelled extensively throughout Britain, aiming to improve his knowledge of farming. As he journeyed, he began to accumulate an unrivalled knowledge of British geology.

In 1767, Hutton moved permanently to Edinburgh, where he became a star of the Scottish Enlightenment, a remarkable period when globally important ideas emerged from this small and relatively poor nation. Hutton's peer group of outstanding intellectuals working and meeting in Edinburgh included the economist Adam Smith (1723–90), the philosopher David Hume (1711–76) and the chemist Joseph Black (1728–99).

It was Hutton's insight into the physical processes of uplift, erosion and sedimentation that convinced him that vast lengths of time were required to form the rocks and landscapes he saw around him. Hutton could see that sedimentary rocks, such as sandstone and conglomerate, were formed from the eroded remains of other rocks, which in turn were formed from the remains of even older rocks. He realized that the landscape had been shaped by very slow, gradual processes that were taking place continuously; 6000 years simply was not enough time to account for all the evidence. Hutton eschewed the textbooks of his day and instead 'read' the rocks and landscapes that surrounded him. Developing and testing his ideas with field-based observations, particularly in Scotland, was the key to Hutton's breakthroughs.

Hutton used many Scottish locations – including Glen Tilt in Perthshire, the Isle of Arran, Jedburgh in the Borders, and Galloway in the southwest – to hone his ideas and convince others of his theories. One noteworthy discovery took place at Siccar Point on the East Lothian coast. Here, in 1788 in the company of the mathematician John Playfair (1748–1819) and the geologist Sir James Hall (1761–1832), Hutton found what he had been looking for: a geological unconformity. At Siccar Point, underlying rocks have been folded, tilted up on end and eroded, and then covered by rocks layered almost horizontally on top. To allow for this feature to form, there had to be a significant break in time separating the two rock types. This understanding was central to Hutton's ideas.

The rocks and landscape of Edinburgh, dominated by Arthur's Seat, Salisbury Crags, Carlton Hill and the Castle Rock, were a source of inspiration for Hutton. There is a commemorative plaque and rock garden where Hutton's house once stood on St John's Hill, from where he had a wonderful view of Salisbury Crags. Volcanic eruptions more than 300 million years ago, followed by deposition of sandstones and then large-scale erosion (most recently by glaciers during the past 2 million years), created the distinctive Edinburgh skyline. The Crags were formed later by molten rock that was squeezed in between layers of sandstone. Hutton answered some fundamental questions about the nature of igneous (once molten) rocks by studying them in detail. One part of the Crags, named Hutton's Section, is now a destination for geologists from all over the world.

In 1783, Hutton was a founding member of the Royal Society of Edinburgh (RSE). He presented the first full paper on his ideas to the RSE in 1785, and his famous book, *Theory of the Earth*, was published three years later. In this book Hutton concluded that he saw 'no vestige of a beginning, no prospect of an end' to the history of Earth. His ideas were retold in 1802, with what most people agree was much greater clarity, by his friend John Playfair, in his *Illustrations of the Huttonian Theory of the Earth*.

Thirty years later, another Scottish geologist, Charles Lyell (1797–1875), developed Hutton's ideas of gradual change into his own concept of uniformitarianism, often summed up in the phrase 'the present is the key to the past'. Lyell's book *Principles of Geology* influenced the palaeontologist Gideon Mantell and Charles Darwin. In fact, Hutton himself had recognized the possibility of the evolution of living creatures in the same way as he saw the evolution of the planet. By opening up the vastness of geological time, James Hutton laid the foundation of modern palaeontology and our understanding of the evolution of life.

Siccar Point on the East Lothian coast shows older rocks tilted almost vertically and overlaid by much younger, near-horizontal layers of red sandstone. James Hutton identified this as a geological unconformity and realized that a significant amount of time must have been needed to form the feature.

Achanarras Quarry

THE BEST-PRESERVED MIDDLE DEVONIAN FISH FOUND ANYWHERE IN THE WORLD

Achanarras Quarry in Caithness, in northeast Scotland, is synonymous with outstanding fossil fish. Some of the world's best-preserved and most diverse fossil fish from the Middle Devonian period have been discovered here. The Devonian is sometimes called the Age of Fishes because during this time, 416 to 359 million years ago, fish were the dominant vertebrates and many new types of fish evolved.

Achanarras is prone to wet and windy weather even at the height of summer, so fossil-hunters should pick their day wisely. However, it is certain that, whatever the weather, the fossils at Achanarras are well worth the trip. The quarry is now abandoned, but was worked for about a hundred years for thin slabs of rock used locally for roofing. The main pit is filled with water but there are mountains of spoil surrounding it. Fossil-hunters are welcome to sift through the piles, providing they follow the Scottish Fossil Code. The fossil fish are found sandwiched between the fine layers of rock and, although anyone can have a go, it is a skilled job selecting the right slabs and splitting them open to reveal the stunning fossils.

An amazing 15 different genera of fish have been found at Achanarras, ranging from the primitive jawless fish *Achanarella* to armoured fish such as *Coccosteus*, *Homosteus* and *Rhamphodopsis*; lobe-finned fish like *Osteolepis*; strange, enigmatic fish such as *Palaeospondylus*; spiny fish including *Mesacanthus*, *Diplacanthus* and *Cheiracanthus*; lungfish such as

Dipterus; and bony ray-finned fish such as *Cheirolepis*. *Cheirolepis* is a distant ancestor of all the bony fish that are alive today, including the brown trout currently living in the quarry loch. A comparison between one of these trout and a fossil specimen of *Cheirolepis* reveals an astonishing similarity in form between the modern-day fish and its 380-million-year-old relative.

Some of the Achanarras fossil fish were scavengers, while others were predators. The lobe-finned fish

Glyptolepis was a top predator and grew to over 1 metre in length. It probably lurked in the shallows waiting for its prey to swim past, like a modern-day pike. One fossil specimen found at Achanarras shows a *Glyptolepis* with an ancient case of greed: a big fish had choked to death trying to swallow a smaller individual of its own kind and had been fossilized with half a fish sticking out from its mouth.

Pterichthyodes milleri was a strange-looking armoured fish, or placoderm, found and named after Hugh Miller (see p. 63), the nineteenth-century Scottish geologist who first studied it. Placoderms thrived from late in the Silurian period (443 to 416 million years ago) to the end of the Devonian period, when they became extinct. *Pterichthyodes* had hard, bony plates covering its head and trunk, eyes on top of its head and a flat belly. Because the fishlike tail is often missing from fossil specimens and there are two big, paddle-like pectoral fins on either side of the box-shaped body, *Pterichthyodes* was initially mistaken for

Today, Achanarras Quarry in northeast Scotland is disused and partially flooded. Many of the broken slabs in the spoil heaps contain beautifully preserved fossil fish, although it takes patience and skill to decide which ones to split open.

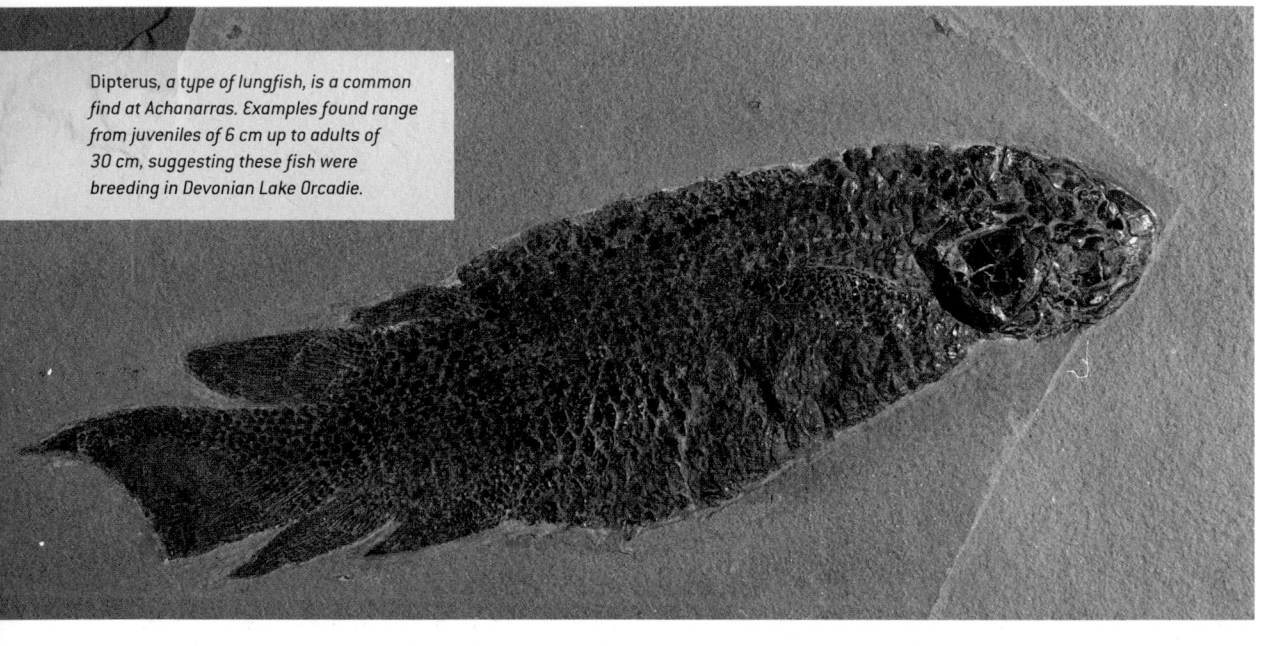

Dipterus, *a type of lungfish, is a common find at Achanarras. Examples found range from juveniles of 6 cm up to adults of 30 cm, suggesting these fish were breeding in Devonian Lake Orcadie.*

a primitive turtle. It was not until fossils were found showing *Pterichthyodes*' scaly tail in place that the fish's true identity was realized.

All the fossilized fish found at Achanarras Quarry lived and died in a vast freshwater lake called Lake Orcadie, which covered the region roughly 380 million years ago. During the Devonian period, most of Britain was part of a much larger landmass that was south of the Equator, with the Devonian Sea covering southwest England. Between mountains to the northwest and the sea was a vast desert plain with its lowest topographic levels occupied by Lake Orcadie, which extended through Shetland, Orkney, Caithness, the Moray coast and western Norway. The Devonian climate fluctuated between dry and wet periods. At times the lake was an oasis in an arid landscape.

The lake teemed with fish, particularly out towards the margins in the shallower, oxygen-rich water. It is not only the fossils of fully grown, adult fish that have been discovered at Achanarras. Fossils of many fish have been found ranging from a few millimetres up to 50 cm long or more, so it can be deduced that fish were breeding in the lake, not just in the rivers and streams feeding it. When the fish died, they drifted out towards the deeper water, where they sank to the bottom and were buried under lake sediments. Year after year, these sediments built up and were eventually compressed and compacted to form the thinly laminated siltstones and flagstones that dominate the geology of Caithness, Orkney and parts of Shetland today.

Hermione and Professor Nigel Trewin of Aberdeen University examine some examples of Devonian fossil fish at Achanarras Quarry. Professor Trewin is an expert on the site.

Some layers of the quarry have abundant well-preserved fish – more than ten individual *Dipterus* have been found in a single square metre, for example – indicating that sometimes many fish died at the same time in 'mass mortality' events. These events were probably due to sudden drops in oxygen levels caused by algal blooms or storms that mixed up the anoxic, or oxygen-free, deep water of the lake with the fresh surface waters. Another possibility is that the lake periodically became very salty due to long droughts and high evaporation. These mass mortalities were clearly a disaster for the fish but a blessing for palaeontologists.

There are several other noteworthy sites around northeast Scotland – such as the coast near Thurso and Helmsdale – where excellent fossil fish have been found, but by far the widest variety have been collected from Achanarras. However, there is one notable absence from Achanarras: *Gyroptychius*, a fossil fish

that has been found in a comparable site in Orkney, which suggests it must have lived in Lake Orcadie. Fossil-collectors at Achanarras should look out for *Gyroptychius* and report any discoveries.

BIOGRAPHY: **HUGH MILLER**

Scotland has been famous for fossil fish ever since Hugh Miller (1802–56) wrote about them in his most successful book, *The Old Red Sandstone*, in 1841. Miller was a gifted scientist and dedicated churchman who became a well-known social commentator.

Some of Miller's collection of more than 6000 fossils is on display at the house where he was born in Cromarty in the Highlands, now a museum. Many of his fossils are in the National Museum of Scotland in Edinburgh, but there are usually at least one or two Devonian-aged fish from Scotland in every large museum in Britain, if not the world.

This specimen of Pterichthyodes milleri *from Achanarras Quarry is complete barring a detached head.* Pterichthyodes *was an armoured fish, or placoderm, with large, flipper-like pectoral fins resembling a pair of wings.*

Clashach Quarry

ANIMAL TRACKS IN THE PERMIAN DESERT

Clashach Quarry near Elgin in Morayshire produces some of the most beautiful building stone in Britain. The Hopeman Sandstone found here clads the nearby Elgin Museum and the new National Museum of Scotland in Edinburgh. Further afield, it has also been used in parts of Gaudi's Sagrada Família cathedral in Barcelona. The sandstone is made from naturally cemented sand dunes that covered much of northern Scotland during the Permian period, about 260 million years ago, when the climate was hot and dry. Living in these harsh desert conditions were animals that walked around in the sand and plodded up and down the dunes. We know about these creatures because hundreds of their fossilized footprints have been found preserved in the sandstone.

Many fossil tracks, all made by four-footed animals, have been rescued with the help of the Clashach Quarry staff. Individual footprints range from just 5 mm to 240 mm across, so we know the size of the animals varied enormously. More than half the tracks have grooves running between the lines of footprints: these are tail drags formed when the animal has dragged its tail through the sand. Some tail drags are intermittent and curve from side to side. In these cases, the footprints also have a distinctive pigeon-toed gait, suggesting that these animals swished their tails as they walked. Some sets of prints are very well preserved, indicating the sand was wet with morning dew when the animals walked on it. Others look a bit

RIGHT: The open pit of Clashach Quarry near Elgin produces building stone of varying colour, ranging from pink and orange to cream, grey and white in different areas of the quarry.

BELOW: A line runs diagonally across this example of reptile tracks in Hopeman Sandstone. This is a tail drag which, as the name suggests, was formed as the animal dragged its tail along the ground, forming a groove that was fossilized alongside its footprints.

squashed and have a pile of sand fossilized behind each print, suggesting the animal was walking uphill, pushing little mounds of sand down the slope as it climbed. Many of the tracks appear to indicate that the animals were walking north, perhaps in search of water in a semi-arid river or lake that is thought to have existed in the centre of the modern Moray Firth Basin at the time.

But what were these animals? Unfortunately it is almost impossible to identify them from their tracks alone, and desert conditions are notoriously bad at preserving body fossils such as bones or teeth. No body fossil had ever been found in Hopeman Sandstone. In the nineteenth century, the famous Elgin Reptiles *Elginia*, *Gordonia* and *Geikia* were identified in a different sandstone not far from Clashach. As the two different rocks are thought to be similar in age, these large reptiles have always been contenders for making the prints at Clashach, but that has never been proved. However, in 1997, something very exciting came to light. The quarrymen at Clashach kept aside a block of red stone with an unusually shaped hole in the side of

it. They showed it to local geologist Carol Hopkins, who was suitably intrigued. Carol alerted Dr Neil Clark, curator of geology at the Hunterian Museum at the University of Glasgow, and he confirmed her suspicions that the small hole could be the entrance to a bigger void inside the rock, known as a fossil mould.

Hermione and Neil Clark from the Hunterian Museum in Glasgow hold the block of Hopeman Sandstone with the unusually shaped hole in the front. The hole was revealed to be the entrance to a fossil mould.

Moulds are formed when bones are encased in sand or sediment that turns to rock; the bone then dissolves away, leaving an empty space in the shape of the bone. Working out what a fossil mould represents can pose problems for palaeontologists. Traditional investigative methods involve filling the void with liquid rubber, letting it set like jelly in a mould, then splitting the rock open to obtain a rubber cast of the original bone. This is

An MRI scan generally gives a better resolution than a CT scan, but previous attempts to use it on fossils had not been successful. In this case, however, MRI scanning gave excellent results. All the voids in the rock had to be filled with a special fluid to show up on the scan. This meant drilling a small hole to open up the sealed chambers and immersing the block for two weeks to ensure every last air bubble had escaped. The final scan took four hours to complete and produced images to 1-mm accuracy. Rapid laser prototyping was then used to

LEFT: *This model of a fossil void was created by MRI scanning of the 'Elgin Marvel'. It revealed a Dicynodont skull. One of the Dicynodont's tusk-like teeth can be seen protruding from the upper jaw to the left of the model.*

BELOW: *This reconstruction of a Dicynodont is based on fossils from Clashach Quarry and elsewhere.*

the way the original Elgin Reptiles were identified, but the technique all but destroys the original fossil. Neil and Carol were keen to avoid any damage to their special block, so they decided to try something new.

Few fossils end up in hospital, but this one was referred to both the Western and Royal Infirmaries in Glasgow for medical scans. Computed tomography (CT) scanning and magnetic resonance imaging (MRI) were used to build up a picture of what was inside the rock, in the same way that they are used to investigate human bodies. The output of the initial CT scan was over 150 images, each representing a 3-mm virtual slice through the stone block. From this, Neil and his colleagues could see that there were additional voids inside the rock, not just those that were accessible from the opening on the side.

The Elgin Museum is clad in Hopeman Sandstone from Clashach Quarry. The museum is home to the 'Elgin Marvel' and examples of fossilized reptile tracks.

The scans revealed that the void in the sandstone was in the shape of the skull of a mammal-like reptile called a Dicynodont. It was identified by the two tusk-like teeth hanging behind the snout. Dicynodonts were herbivores, with beaks like a tortoise, and grew up to 2 metres long. They were reptiles but were related to mammals, not dinosaurs or crocodiles.

create a physical plastic model of the fossil void based on the images from the scan. It was the first time that this combination of technologies had been used, and most importantly the block was left completely intact and undamaged, ready for further study in the future.

For the first time after more than 150 years of looking, there was direct evidence of which creatures had made at least some of the fossilized footprints in the Hopeman Sandstone. The newly discovered Dicynodont resembles *Gordonia*, one of the original Elgin Reptiles found in the region, but the one from Clashach is bigger and a different species. Given the range of footprints found, it is likely there were many Dicynodonts and other reptiles in the area. Hopefully more fossil discoveries will build up a more complete list in the future.

The extraordinary hole-in-the-rock fossil dubbed the 'Elgin Marvel' can be seen on display at the Elgin Museum. The reconstruction of the skull and some fossil footprints are also on display, in a tribute to one of the most exciting fossil finds of recent times.

The Rhynie Chert

THE OLDEST PEAT BOG

One of Scotland's most important fossil localities is completely hidden below the gently rolling fields and hills of Aberdeenshire. It remained a geological secret until its chance discovery by a local doctor in 1912.

An astonishing 408 million years old, the Rhynie Chert (a legally protected SSSI) preserves the earliest known bog community, with its tiny plants and animals trapped in hard silica rock. So rapid was the process of fossilization that this flora and fauna was preserved in miraculous detail. The delicate tissues and cells of the primitive plants are so well kept that some experts were originally sceptical about their extraordinary age.

In early Devonian times, a peat bog developed around a hot spring in whose waters primitive land plants grew. Few of them reached heights of more than 20 cm and, like the living primitive plant *Psilotum*, these taller plants were simple branched, but leafless, vertical shoots with club-shaped reproductive structures. Known as tracheophytes, the stem cells of these plants were internally strengthened to help them grow upright. Even more primitive land plants, such as lichens and mosses, lack these structures and cannot grow upright. Detailed microscopic study of the Rhynie rocks shows that several kinds of primitive plants had already evolved by early Devonian times, including *Asteroxylon*, an early clubmoss (lycophyte), some of whose Carboniferous descendants were to grow into 30-metre-high trees.

The Rhynie bog plants were also home to a variety of tiny arthropods, most of which were only a few millimetres long. They include crustaceans such as the fairy-shrimp-like *Lepidocaris*; myriapods with

Astute detective work by a local doctor led to the discovery of one of the most important sites for primitive land plants in the world, near Rhynie.

predatory and detritus-eating forms; clawed (chelicerate) spider relatives known as trigonotarbids, such as *Palaeocharinus*, which was carnivorous; the world's oldest mites, such as *Protacarus*, which fed on dead organic matter; springtails (collembolans) such as *Rhyniella*; and some extinct crustacean-like euthycarcinoids.

the carnivorous arthropod predators. Since no fish appear to have lived in these bog waters, it seems that there was no connection to any river system.

The Rhynie Chert was discovered by chance in 1912 when a local doctor and keen amateur geologist, Dr William Mackie, found lumps of the chert in a field near the village of Rhynie. Spotting the curious

A slice of Rhynie Chert reveals plant stems whose details identify Asteroxylon, *one of the most complex of the primitive plants found here.*

Plant material is difficult for animals to digest, and this community of ancient plants and animals helps us to understand how animals first adapted to plant-eating. As the plants died, their tissues were broken down, by bacteria and fungi, to form a boggy decaying mat. The organic debris was more digestible for some of the myriapods and mites. While feeding on the organic detritus, they also ingested the bacteria and fungi. These micro-organisms took up residence in the animals' guts, and eventually such gut floras were used to break down fresh plant material. At the top of the food chain were

fossils, he used his technical expertise to prepare thin sections of the rock and view the fossils with a microscope. Realizing their importance, he alerted some academic specialists to what he had found. Later that year, following clever geological detective work, a Geological Survey fossil collector, D. Tait, had a trench dug to expose the chert strata and collect more samples. Further trenching in the 1960s and 70s provided yet more material that is still revealing new treasures from the Rhynie bog.

The north of England offers a wealth of opportunities for eager fossil detectives. Many enthusiasts head straight to the excellent fossil-hunting grounds on the Yorkshire coast. Here, the rocks get progressively younger as you head south, from exposures of lower Jurassic shales at Staithes, to dramatic Cretaceous Chalk cliffs at Flamborough Head.

With easy access to beach sites such as Runswick Bay, there is a high chance that you will find something, particularly Jurassic ammonites, and maybe some beach-washed jet. Although rare, it is also worth keeping your eyes open for a large vertebrate – in July 2007, an ichthyosaur skeleton was found in the foreshore near Saltburn. High rates of cliff erosion and tidal activity ensure a constant supply of fossils on the beaches, but some parts of this coastline can be treacherous, so it is best to join a guided fossil walk if you are new to the area.

Away from the coast, northern England is dominated by Carboniferous rocks. Carboniferous literally means 'coal-bearing'. Thick seams of coal are sandwiched between layers of sandstone and shale, and have played a central role in the economic fortunes

At Saltwick Beach on the North Yorkshire coast, the rocks in the cliffs date from the Jurassic period, about 170 million years ago, and some sandstone layers contain fossilized footprints of dinosaurs that once roamed the area.

of the region, as well as offering an insight into life at the time when the coal was formed.

Over on the west coast, there is a chance to get a fleeting glimpse of the activity of our ancestors. On the beach at Formby, dating back almost 5000 years, are footprints preserved in mud. They are not old enough to be fully fossilized, but they are a fascinating insight into prehistoric life (see p. 78).

Jet

FOSSILIZED WOOD FROM NORTH YORKSHIRE

For thousands of years, jet has been used for jewellery. It is often thought of as a semi-precious gemstone but is actually a particular type of fossilized wood. The history of jet is inexorably linked with Whitby on the North Yorkshire coast. The fortunes of the town have been closely tied to the rise and fall of a jet-mining industry, and the surrounding landscape bears the scars from a time when jet was mined by hundreds of people. Jet is not unique to the Whitby area — small amounts can be found in other parts of Britain and abroad — but the rocks around Whitby are particularly rich in hard, dense jet, which is ideal for jewellery. Today, you can still successfully hunt for pieces of this unusual fossil that have been eroded out of the cliffs and washed up on the beaches of North Yorkshire, just as people have been doing there for millennia.

The first thing you notice about a lump of jet is how light it is. It is also lustrous, durable and has an intense, velvety-black colour, all of which explains its appeal as a semi-precious 'stone'. Whitby jet is formed from a species of ancient *Araucaria* tree very similar to the modern-day Monkey Puzzle. Under a powerful microscope, the fossilized tree rings and cell structures of the original wood are still visible. The jet occurs as small lumps and thin seams in rocks that are about 185 million years old. Back then, *Araucaria* trees flourished in forests alongside the dinosaurs that lived there. It is thought that the characteristic spiky branches of these trees may have been an adaptation to prevent predation by giant plant-eating sauropods.

The jet-forming process began when dead or uprooted trees were swept out to sea by rivers or floods. Saturated

with seawater, the tree trunks and broken-up logs sank to the bottom, where they were compressed and fossilized between layers of anoxic (oxygen-free) marine sediments. Those marine sediments now form the jet-bearing shales of North Yorkshire that underlie the North York Moors and crop out at various locations along the coast from Ravenscar to Saltburn. Made from compressed, organic matter and with a high carbon content, jet is essentially a form of coal, but unlike coal, true jet is found only in marine rocks.

Jet has been treasured throughout history. Talismans and jewellery carved from it have been found at archaeological sites dating back thousands of years. Intricate jet-bead necklaces, thought to denote high status, have been found in early Bronze Age burial sites in Britain. The Romans used Whitby jet for amulets, jewellery, dagger handles and even dice. During the Middle Ages, jet was used for religious artefacts, healing and keeping away evil spirits.

Whitby has always been at the centre of the jet trade but, when the Victorians developed a particular passion for jet in the mid-nineteenth century, the industry peaked. Jet was perfect for the Victorian fashion for large, ornately carved brooches and beads because it is

As jet is lightweight, it was ideal for the Victorian fashion for big jewellery, like this necklace, earrings, brooch and pendant carved from Whitby jet c. 1870.

Pieces of beach-washed jet found near Whitby. The larger piece is 3 cm across. Note the rounded, worn edges and slightly pitted surface typical of jet that has tumbled about in the sea.

lightweight and takes a very high polish. After the death of Prince Albert in 1861, Queen Victoria took to wearing jet jewellery exclusively, and as a result it became even more popular. Such was the demand for Whitby jet that the industry grew from just two shops with a handful of employees in 1832 to a flourishing trade with more than 200 shops and 1500 workers in 1872.

Before the Victorian era, the jet industry had been supplied with fragments collected from beaches and natural cliff falls, but as demand grew people began mining jet by tunnelling into the cliffs and digging into the shales across the North York Moors. Good-quality jet is found in a relatively brief 'window of time' in a succession of lower Jurassic rocks called the Mulgrave Shale Member. The miners used a thin layer of limestone called the Top Jet Dogger as an upper marker to help identify where to find the Mulgrave Shale.

The Jet Dogger caps a layer of shales about three metres thick. The best-quality jet is found here, but seams are usually very short and quickly exhausted because they are formed from isolated trees and logs. This meant that the Yorkshire jet-mining industry was very labour-intensive. By the beginning of the twentieth century, cheaper jet from Spain and imitation products such as vulcanite had made jet mining in Yorkshire unprofitable. As fashions changed, and jet jewellery became less popular, the Whitby jet industry went into rapid decline.

Although it is not as popular as it was in Victorian times, the past few years have seen a resurgence of interest in Whitby jet, and the streets of the town are once again filled with shops selling jet jewellery and ornaments. One of the current crop of Whitby jet carvers is Mike Marshall, who collects, carves and polishes jet using traditional techniques, but is also an expert in identifying and preparing other fossils that he finds along the coast.

As Mike explains, the trick is to walk the beach, keeping your eyes peeled for small, rounded, shiny black lumps that may be up to a few centimetres across. However, there are many other black pebbles and little bits of brittle coal littering the beach, and distinguishing a piece of jet is not straightforward. Like all fossil-detecting, finding jet takes a keen eye and a bit of luck, but if your patience runs out there are plenty of shops in the town selling fossil jewellery.

A modern Monkey Puzzle tree (Araucaria araucana) growing in its native habitat in South America. Jet is formed from a similar type of tree that grew 185 million years ago.

Triassic Deserts

PLANTS FROM THE WIRRAL AND THE MYSTERY OF THE CHIROTHERIUM

The eye-catching red sandstone rocks seen in the Wirral and Liverpool area were formed from ancient sand dunes that covered almost the entire region during the Triassic period, around 240 million years ago. At this time, Britain was part of a supercontinent called Pangaea and much closer to the equator, at about 30° north. The rocks conjure up an image of a hot and barren desert, but fossil evidence from the area reveals that, at times, there was a surprising amount of animal and plant life around. It seems that the Triassic was not always as dry and desolate as its popular image, and there were some wetter periods more conducive to life.

Sandstone exposures in the road cuttings around the Wirral Country Park, like this one at Grange Hill, allow you to 'look inside' the ancient dunes that formed the rocks. Individual sand grains have a smooth, regular appearance indicative of aeolian (wind-blown) deposits. The diagonal lines, known as cross bedding, are typical of sand dunes. The red colour is due to high levels of iron oxide, another indication of dry conditions when the rocks formed.

The Wirral has been famous for fossilized footprints in Triassic rocks since the early nineteenth century, but identifying exactly which animals made them has proved difficult. Some of the tracks, originally discovered in Storeton Quarry in 1838, bear a striking resemblance to human handprints. The tracks were named *Chirotherium*, which literally means 'hand beast'. It took more than 130 years for the tracks to be correctly identified as belonging to meat-eating reptiles called rauisuchians (see *The Chirotherium Mystery* on p. 77). One reason it was hard to identify the *Chirotherium* track-maker was a lack of fossil bones. If prevailing conditions were hot and dry, skeletons were prone to crack and break up in the heat before they got a chance to fossilize. However, if the ground was periodically damp, if only for a few hours, freshly made footprints would fossilize more easily after being dried out and baked hard. This may explain why so many tracks have been found despite a complete lack of fossil bones or teeth.

As well as *Chirotherium*, footprints belonging to tortoises and other reptiles called rhynchosaurs have also been identified in the area. Rhynchosaurs had long, thin toes and claws, and their footprints look a bit like scratch marks in the rock. It is likely that the meat-eating rauisuchians would have survived by preying on the smaller rhynchosaurs. But that does not explain why the numerous rhynchosaurs, who were herbivorous, lived in this particular place. Up until recently there was very limited fossil evidence for ancient plant life in the region. But a new discovery by scientists from the World Museum Liverpool shows that there could have been many more plants growing in this particular area than previously thought. The rhynchosaurs probably survived on an abundance of plant life after all.

ABOVE: *Hermione and Alan Bowden, in the laboratory at the World Museum Liverpool, examine fossil plant fragments from Triassic-aged rocks in the Wirral.*

BELOW: *One of the Wirral plant fossils found by Bowden and his colleagues bears a strong resemblance to the tip of a modern-day horsetail (inset). Three distinct types of horsetail have so far been identified.*

ABOVE: A drawing of a rauisuchian reptile of the type thought to have made the Chirotherium tracks.

RIGHT: A slab of sandstone from the Wirral shows two very distinct fossilized Chirotherium tracks. Note the similarity to human handprints, with the 'thumbs' on the outside of the prints. Other markings on the slab that look like scratch marks were left by smaller reptiles called rhynchosaurs.

In 2006, in the Wirral Country Park, Alan Bowden and his colleagues Geoffrey Tresise and Wendy Simkiss, from the World Museum Liverpool, found some very rare plant fossils in Triassic sandstones about 240 million years old. By carefully analysing the fossils, Alan Bowden's team are trying to reconstruct the plants that grew in the Merseyside area at the time, but it is far from easy. The fossils are a jumble of tiny plant fragments and might easily have been overlooked by an amateur fossil detective. Imagine trying to recreate the plants in your garden from the contents of your compost bin after it had not been emptied for years. That is how Alan describes the scale of the task they face.

In the Earth Sciences laboratory in the basement of the museum, Alan examines the tiny plant fossils under the microscope. The largest plant fragments are just a few millimetres long, visible as little black flakes of carbon on lumps of sandstone. On closer inspection, one intriguing specimen has recognizable features. It shows the tip of a tiny horsetail, and when compared to an image of a living horsetail the resemblance is unmistakable. As well as three distinct types of horsetail, the team have found fragments of a conifer probably related to an early form of spruce, and other as yet unidentified plant fossils that they think might be lycopods such as *Pleuromeia*, soft-leaved ferns and Cycadalean plants. As well as identifying the plants, Alan and his team are trying to work out why the fragments are so small. They are investigating what happens when examples of modern plants related to the fossil specimens dry up and begin to disintegrate.

The World Museum Liverpool's work on plant fossils is one piece of the palaeontological jigsaw that explains life in the Wirral during the Triassic period. In time, they should get a much clearer picture of Triassic plant communities and, significantly, the role plants played in supporting the varied animal life. The research already confirms that, during the generally dry conditions of the early to middle Triassic period, there were wetter periods capable of supporting a variety of plants. A potential modern-day analogue for the environment of the Wirral 240 million years ago is the river valleys of northern Chile. Here, fertile pockets of land are inhabited by monocultures of horsetails and conifers. The valleys are home to indigenous reptiles, too, while at the same time they are surrounded by the hot and arid Atacama Desert.

A range of Triassic reptile footprints and reconstructions of the animals that formed them can be seen at the World Museum Liverpool. The surrounding area is full of clues, too. Storeton Quarry is filled in now, but near to where it used to be in Higher Bebington, you can see *Chirotherium* prints on a sandstone slab built into the porch wall of Christ Church. Recently, new fossil footprints have been found on Hilbre Island near the mouth of the River Dee — more evidence of the reptiles that roamed the North of England millions of years ago in the desert and not-so-desert conditions of the Triassic.

DID YOU KNOW? **THE CHIROTHERIUM MYSTERY**

The *Chirotherium* mystery is a classic palaeontological detective story. It began when the very first hand-like *Chirotherium* fossils were found in Germany in 1833. Scientists speculated about what type of creature could have made the footprints, and initial suggestions included an ape, a bear, a marsupial and a giant toad. For a long time, following a suggestion by Richard Owen in 1842, it was thought that a large, extinct amphibian known as a *Labyrinthodont* was responsible, but none of these early ideas proved to be a suitable match. Throughout the nineteenth century, the hunt continued for a creature with two large hind feet about 35 cm long, each with four forward-facing toes and a sticking-out 'thumb', and two smaller but similarly shaped front feet. Such was the likeness to human hands that for many years scientists were convinced that it was definitely a thumb, or first digit, that was sticking out rather than a little toe or fifth digit. Because the 'thumb' was on the outer edge of trackways, they thought the unknown creature had walked with its legs crossed.

Hundreds of *Chirotherium* prints were found at Storeton Quarry and the surrounding area in the northwest of England. When quarrymen first saw the prints, they thought they might have been made by victims of Noah's Flood, reflecting popular understanding of geology at the time. *Chirotherium* in Britain are almost always found as fossil casts of the original footprints, so they protrude from the rock surface rather than forming depressions. They were formed when wind-blown sand filled in the original footprint moulds in a layer of soft clay. When quarried, the rocks naturally split apart along the weaker clay layers to reveal *Chirotherium* prints in relief on the underside of sandstone slabs.

Despite the numerous discoveries in England and in Triassic-aged rocks right across Europe, from Italy to Spain, during the nineteenth century, still no fossil remains of a suitable animal were found. By the beginning of the twentieth century, scientists were well aware that reptiles had been the dominant land animals in the early Triassic and, in 1925, a reptilian contender was identified in South African rocks of the right age. Bones of a rauisuchian reptile (ancestral to the dinosaurs) called *Euparkeria* appeared to have the characteristic jutting-out fifth toe, but the mystery still was not over because *Euparkeria* was too small.

At last, in the 1960s, bones of a rauisuchian called *Ticinosuchus ferox* were discovered in Switzerland — the right age, the right size and with the right-shaped back feet. Scientists are confident that the case of the *Chirotherium* has been solved: it was a closely related *Ticinosuchus* species that lived in the Wirral 240 million years ago, and left behind its mysterious footprints.

The Ephemeral Footprints

AT FORMBY SANDS

The beach at Formby Point in Lancashire is a long and largely unspoilt stretch of sand, but occasionally the sea exposes some patches of hard mud in the intertidal zone. Look closely, and one or two of these patches might display some intriguing impressions: a few are unmistakable as human footprints, while others look like the tracks of birds, deer and some very large cows. The imprints seem deceptively fresh, as if they were made only a few days ago, but in fact these tracks were left by our ancestors and various wild animals between 3500 and 4500 years ago.

Gordon Roberts, a retired teacher now in his seventies, first spotted the puzzling prints in 1989 and has been studying them ever since. He knew immediately that they had to be old – for a start, no cow alive has such big feet, and the continuous trackways disappear beneath overlying layers of sediment. For almost 20 years, he and his wife Pat have surveyed the beach in all weathers. He has built up an unrivalled archive of notes, photographs, plaster casts and measurements of more than 150 different sets of tracks, spread along 2½ miles of beach. This is an invaluable database because, as fast as the sea exposes the fossil footprints, it washes them away.

Strictly speaking, the footprints are sub-fossils because they are still soft and provide just a fleeting glimpse of life from a few thousand years ago. Had they remained buried, these footprints could, with time, have become fully fossilized. They were created when people and animals walked across a coastal mudflat, leaving their tracks to be baked hard by the sun. Sand, silt and mud covered the prints, and layers of sediment built up on top. Sea levels have fluctuated over the past few thousand years, and it is a lucky coincidence that the present-day beach is aligned with the area of footprints and that erosion has removed just enough of the overlying sediments to make them visible.

The beach at Formby Sands on the Sefton coast in Lancashire. Layers of hard mud beneath the sand are often exposed by the sea, revealing human footprints made by our ancestors about 4000 years ago.

Gordon has recorded evidence of humans (both adults and children), roe and red deer, horses, wild boar, dogs/wolves and cranes. He also has evidence of aurochs – extinct wild cattle much larger than modern cows – which would explain the very large hoof prints. What you can see on any one day is dependent on the tides and the wind, but if you are lucky, at low tide, you might see some very clear human footprints. Although these footprints may look as if they were made recently, you will soon discover that your own feet make no impression at all on the hard mud despite its being wet and slippery underfoot. On many prints, the outlines of individual toes can be traced. It is a sad sight to see the footprints covered by the incoming tide; in just a few days they will be gone for ever.

Gordon works closely with a group of scientists, including Professor Matthew Bennett, a human

One of the hundreds of ancient human footprints that have been found on the beach at Formby. It may look fresh, but it was created when mud hardened soon after the impression was first made. Some prints show evidence of missing toes and bunions.

For the past 20 years, Gordon Roberts has monitored the beach at Formby as often as possible, building up an unrivalled record of the prints left behind by prehistoric humans and animals.

footprint expert from Bournemouth University, who is using laser technology to take very detailed 3D digital scans of the prints. Computerized scans are superior to casts or photographs because they can be manipulated and compared much more efficiently. Matthew can only bring the scanning equipment to the beach occasionally, but Gordon keeps a daily record of what is exposed. Matthew explains that the Formby footprints are only as well documented and understood as they are because of Gordon's unprecedented dedication to the project.

By collecting data on lots of prints, Matthew hopes to find out about the population structure of the people living in the area at the time and to learn about their health and lifestyle from their feet. It is possible that both people and animals were using the foreshore for easy access up and down the coast because inland the route was either too densely wooded or waterlogged. The mudflats could also have been used for hunting and gathering because it was an area rich in food, such as plants, shellfish, fish, wildfowl and mammals. The shape and size of the human prints can reveal information about height, weight, gait and pace, but whether they show children playing, women gathering shellfish or men running after deer is a moot point. After all, a tall woman and a short man might leave very similar prints.

The Formby footprints are among the youngest evidence for ancient life considered in this book but they are an important part of a sparse global database of prehistoric human footprints. In Britain there is just one other comparable (intertidal) site on the Severn Estuary, less extensive than Formby and very slightly older. More sites like these may yet be found if the right person, like Gordon Roberts, happens to be on the right beach at the right time. In the meantime, it is important to gather all the information that we can about the ancient animals and humans from Formby before the beds are eroded – and there is no time like the present. Gordon calls his work 'ephemeral archaeology' because, as fast as his lost world of Formby Point is revealed, the sea destroys it for ever.

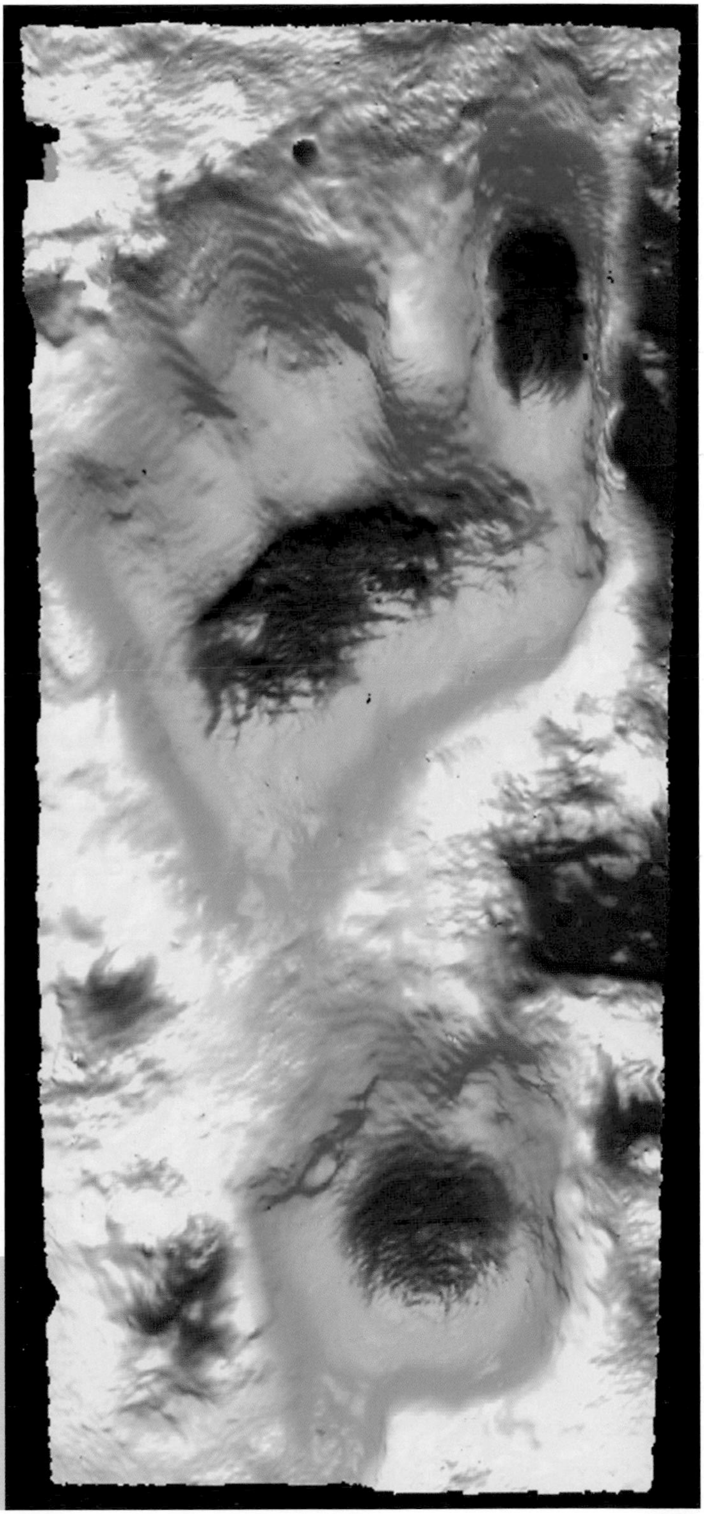

A colour image of a 3D laser scan of one of the Formby footprints made by Matthew Bennett from Bournemouth University. The scans reveal many details of the shape of the prints and can be easily compared with prints from other locations.

William Buckland

STRANGE CAVE FOSSILS IN YORKSHIRE

Historically, one of the most famous cases of fossil detective work was carried out by the Reverend Doctor William Buckland, reader in mineralogy at the University of Oxford, on a cave at Kirkdale in Yorkshire. The cave, of Carboniferous limestone, was discovered by quarrymen in the 1820s. Its floor was covered in bones.

Dr Buckland was particularly interested in cave fossils as part of his defence of the Old Testament as a historical account of creation and events such as the Flood. His investigation of Kirkdale Cave revealed the remains of over 20 different species of mammals, including some that were native to Britain (for example, deer species), mixed with an exotic assemblage of big cats, bears, rhinoceros, horses and elephant-related species, many of which were far too large to have entered through the relatively small cave opening when they were alive.

Although the bones from different skeletons were all jumbled together, they were so well preserved that Buckland realized the animals must have lived locally, even though some belonged to exotic species. Another explanation was that they might have been carried to Yorkshire by the waters of the Flood. Buckland modified this theory – he argued that the absence of similar bones outside the cave was due to the fact that they had been swept away by the Flood waters. Only the bones in the deep recesses of the cave survived. But how did they get into the cave?

Close examination revealed the presence of tooth marks and fractures similar to those produced by dogs. Buckland also noticed numerous hyena bones, including 300 canine teeth. He knew that hyenas had powerful jaws and were able to crack open even thick bones in the pursuit of nutritious marrow. As luck would have it, around this time, a live hyena from South Africa came to Oxford with a travelling circus. Buckland took the opportunity to watch it feed, noting what it regurgitated and what it defecated. Buckland also imported a live hyena with the intention of dissecting it, but became so attached to the

Buckland, surrounded by fossils, demonstrates features of a fossilized hyena jaw, showing that it was powerful enough to chew bones just as modern hyenas do.

creature that it joined his bizarre family menagerie. Billy, as it was named, survived for another 25 years, along with a bear and a monkey. As Buckland wrote to a friend: 'Billy has performed admirably on shins of beef, leaving precisely those parts which are left at Kirkdale and devouring those that are wanting ... so wonderfully alike were these bones in their fracture ... that it is impossible to say which bone had been cracked by Billy and which by the hyenas of Kirkdale!'

Buckland had also found fossil faeces (called coprolites) in the cave, and on dissection found them to be full of bone fragments. He put all his experimental observations together and reconstructed the original scene. He argued that, as active scavengers, the hyenas had brought individual bones

A contemporary cartoon shows Buckland entering Kirkdale Cave as if it were still occupied by hyenas and the remains of their scavenging activities.

into the relative safety of the cave to chew over at their leisure away from other predators and scavengers. All his results were presented as part of a book published in 1823, entitled *Reliquae Diluvianae; or, Observations on the organic remains contained in caves, fissures, and diluvial gravel, and on other geological phenomena attesting the action of an universal deluge.* For his pioneering detective work, Buckland was awarded the prestigious Copley gold medal by the Royal Society.

Living hyenas have remarkably powerful jaws and the persistence to keep gnawing bones until minute fractures spread and eventually open to allow them to get at the highly nutritious marrow.

The age of the fossils in the following stories from central England ranges enormously, reflecting the diverse geology of this region. The chapter begins with mysterious Precambrian life forms more than 560 million years old and ends with an Ice Age mammoth that was still alive just 12,700 years ago.

In both these cases the initial evidence was found by a member of the public, not a trained palaeontologist – proof that anyone can be a successful fossil-hunter. Other *Fossil Detectives* stories from the region include the world-renowned Wenlock Limestone, featuring the Dudley Bug trilobite, and

some exceptionally well-preserved ancient marine creatures from a secret location in Wiltshire.

Historically, many intriguing fossil finds have come from the Jurassic shales and limestones around Oxford. In 1677, Robert Plot, keeper of the Ashmolean Museum, proposed that a large bone fragment found in a local quarry was part of an elephant brought to Britain by the Romans, or more likely the thigh bone of an extremely large man. Without knowing it, Plot had actually written one of the first descriptions of a dinosaur bone. In 1824, William Buckland, reader in

Charnia

THE OLDEST COMPLEX LIFE FORM ON EARTH

The beauty of fossil-hunting is that anyone can do it. And so it follows that anyone, with a bit of luck, could make the next big discovery. If you spend a bit of time investigating your local geology, the chances are that you will make a discovery, and that discovery – even today – could be of something new that revolutionizes our understanding of the history of life on Earth.

A perfect illustration of this took place in 1957 near Leicester. A 15-year-old schoolboy made a remarkable discovery and inadvertently rewrote the rule book about the emergence of complex life on Earth. Roger Mason and two school friends, Richard Allen and Richard Blachford, spotted some unusual markings and leaf-like surface impressions while rock climbing in an old quarry in Charnwood Forest to the north of the city. It is not surprising that these fossils had lain undiscovered, because no one would have thought to look there. The rocks of Charnwood Forest are Precambrian in age, created by volcanic eruptions and underwater landslides in an ancient sea about 560 million years ago. Rocks of that age simply could not contain any sizable fossils because complex life did not exist back then – or so everyone thought. Simple, single-celled, microscopic organisms, such as algae, were all that had been found in such ancient rocks. Only after the explosion of life at the start of the Cambrian period, about 542 million years ago, were complex creatures thought to emerge. But despite this accepted way of thinking, Roger knew he was on to something.

With the help of his father, he convinced geologist Trevor Ford, from the University of Leicester, to take

Wenlock Edge is a forested escarpment of Wenlock Limestone in Shropshire. Managed by the National Trust, it is one of the best places to see fossilized reef communities preserved in limestone dating from the Silurian period, 420 million years ago.

mineralogy at Oxford University, correctly identified similarly large bones, from Stonesfield Quarry to the north of the town, as belonging to a giant reptile. He called his fossil reptile *Megalosaurus* in what was officially the first published account of a dinosaur. Today, you can see Buckland's *Megalosaurus* at the Oxford Museum of Natural History, which is home to an extensive collection of fossils from Britain and abroad.

This is a rubber mould of a new specimen of
Charnia masoni *found close to where the original
discovery was made in Charnwood Forest in 1957.
It is approximately 25 cm long.*

a look, and persuaded him that these were genuine
fossils. Ford was convinced and published the finds
in the *Proceedings of the Yorkshire Geological Society*
the following year. He named the two different fossil
types *Charnia masoni* (after Charnwood Forest and
Roger Mason) and *Charniodiscus concentricus*
(because of the location and resemblance of the lower
part of this fossil to disc-shaped, concentric rings).
The discovery was certainly ground-breaking, but
as with so many new ideas in science, it raised more
questions than it answered.

For a start, what were these surprising fossils? What
could have been living on the sea floor so long ago,
swamped by volcanic debris and turned to stone?
Charnia masoni resembled a thick, plaited, almost
fern-like frond, but it was unlike any animal alive today.
It grew attached to the sea bed on a disc-shaped hold-
fast or foot, but nobody really knows how it survived.
Perhaps it filtered microscopic matter from the water

for nutrients. There is no evidence of any hard parts
in *Charnia* – they lived well before the development of
internal or external skeletons – so it is fairly miraculous
that we have fossils of them at all. But as it turns out,
the fossils at Charnwood Forest are not alone.

Similar findings had been recorded in Newfoundland in
the 1860s, and later in Namibia and Australia. But little
attention was paid to these discoveries because the
rocks were thought to be younger than Precambrian,
and therefore unremarkable. Until the late 1940s, it
was still thought that Precambrian sedimentary rocks
were virtually devoid of fossils. And so, by definition, the
discovery of any early fossils meant that the strata that
contained them were of Cambrian age. In central England,
however, there was good evidence that the Charnwood
strata lay below Cambrian strata and were genuinely
Precambrian. After the publication of Ford's article,
scientists realized there was a whole fauna of complex
creatures around in Precambrian times.

These marine life forms are known collectively as the Ediacarans, named after the Ediacara Hills in South Australia, where a good collection has been found, including *Charnia* and others such as *Spriggina*, *Dickinsonia* and *Arkarua*. Some resemble discs, mats or quilts, and have been described as 'lettuce-like' or even 'pizza-like'. Ediacarans have now been found in more than 30 sites worldwide, firmly establishing them as a successful, if ultimately doomed, group of organisms living before animals and plants appeared (in the way that we usually think of them).

A great deal of debate still surrounds the Ediacarans, mainly because it is not clear what kind of organisms they were. Fossil detectives from around the world are still searching for missing pieces that may solve this puzzle. The global distribution of the Ediacarans and the fact that they existed for more than 20 million years suggest that they were the first abundant and diverse large organisms in marine strata of the late Precambrian age. They are the first known organisms that can be considered truly multi-cellular.

More than 150 different Ediacaran genera are now known from around the world, with slight but important differences in their ages and the kind of marine environment in which they are preserved. Among the oldest is the spectacular biota found in Newfoundland, which is about 575 million years old. Some of the frond-shaped fossils in Newfoundland are more than 1.6 metres in length. Some of the youngest have been found in Namibia, and they include several sack-shaped forms that lived within the sea-bed sediment. Many of the best-preserved Ediacarans have been found in Arctic Russia, and are about 555–558 million years old. These are proving to be of particular importance in arguments over Ediacaran biology, especially a complex-looking fossil called *Kimberella*, of which 800 specimens are known. *Kimberella* seems to have been a mobile mollusc-like creature and therefore could be an ancestor to some of the animal fossils found in Cambrian strata.

With the possible and much-disputed exception of some cnidarians (members of the phylum cnidaria, which includes jellyfish and sea pens), molluscs, arthropods and worms, no post-Cambrian organisms are thought to be descended from the Ediacarans, which suggests that most of them belonged to completely extinct and as yet unknown groups of organisms. Whatever they were, this discovery has proved that complex life started much earlier than previously thought, and some of the best evidence for these ancient life forms comes from central England.

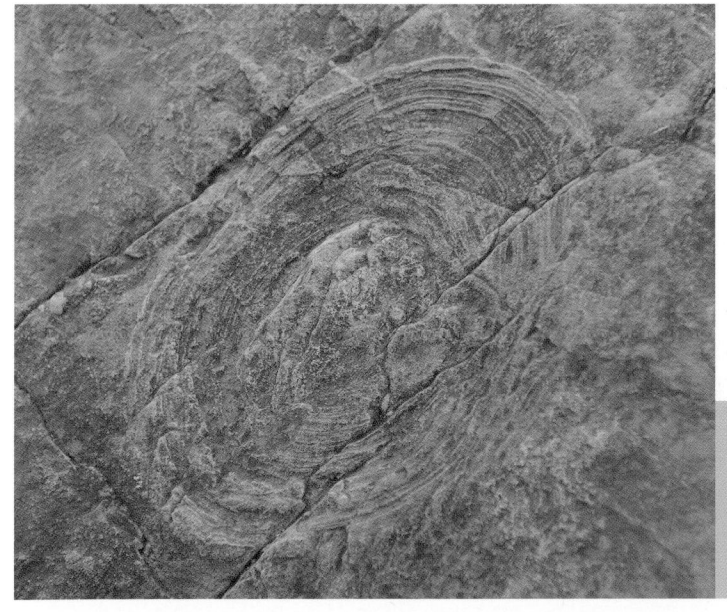

The raised concentric rings of Cyclomedusa davidii, *another Ediacaran fossil found in Charnwood Forest. This specimen is 40 cm across. Several others, some just 2 cm in size, have been found nearby. It may have resembled a jellyfish, although exactly what type of creature it was remains a mystery.*

Wenlock Limestone

... AND THE DUDLEY BUG

Central England is the only *Fossil Detectives* region without a coastline. You might think that, with no freely accessible cliffs or beaches to investigate, it would be more difficult to find productive fossil-hunting grounds. But there are still lots of good places to try and, when it comes to exploring the world-famous Wenlock Limestone deposits in the West Midlands and Shropshire, it isn't hard to find a fossil within minutes.

Wenlock Limestone contains the fossilized remains of marine creatures that inhabited a warm, shallow sea that covered much of England during the Silurian period, about 420 million years ago. Rather than just a few species, whole communities of diverse animals were fossilized together here. During the Silurian, central England was positioned roughly 25° south of the equator, and corals, sponges, bryozoans and

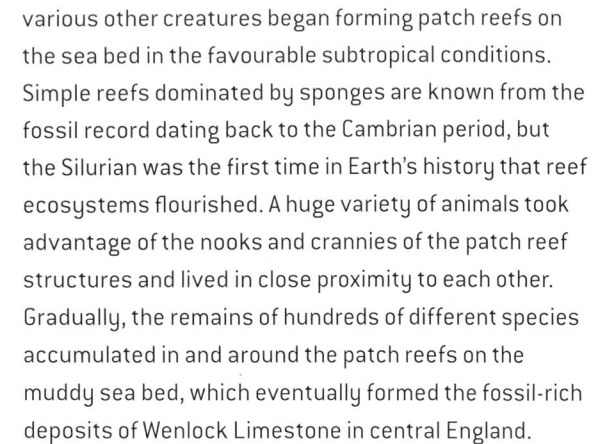

various other creatures began forming patch reefs on the sea bed in the favourable subtropical conditions. Simple reefs dominated by sponges are known from the fossil record dating back to the Cambrian period, but the Silurian was the first time in Earth's history that reef ecosystems flourished. A huge variety of animals took advantage of the nooks and crannies of the patch reef structures and lived in close proximity to each other. Gradually, the remains of hundreds of different species accumulated in and around the patch reefs on the muddy sea bed, which eventually formed the fossil-rich deposits of Wenlock Limestone in central England.

Silurian patch reefs were relatively small, isolated, circular structures compared to a modern reef such as the Great Barrier Reef in Australia. You can still see rounded and domed structures called bioherms, up to a few metres across, in some Wenlock Limestone exposures, which reveal the original structure and shape of the patch reefs. One of the best places to look for them is Wenlock Edge in Shropshire, an escarpment, or ridge, that forms a series of low hills about 330 metres high. The escarpment stretches from the town of Much Wenlock in a southwesterly direction for about 15 miles. This landscape was the inspiration for A.E. Houseman's poem 'On Wenlock Edge the Wood's

The ancient woodland on Wenlock Edge in Shrophire. Underfoot, Wenlock Limestone is teeming with fossils of marine creatures that lived in and around Silurian-aged reefs about 420 million years ago.

The sea bed in Silurian times was home to a wide variety of animals, including corals and crinoids, that lived in early patch reef systems. Wenlock Limestone formed from the compressed remains of this type of sea-bed community.

in Trouble ...' in his celebrated book *A Shropshire Lad*, published in 1896. The area is managed by the National Trust and there are marked walks through the ancient woodland on the ridge, but it is the rocks underneath, absolutely teaming with fossils, that are the real attraction for fossil fans.

Some of the most common fossils to be found in Wenlock Limestone are tabulate corals such as *Favosites* and rugose corals such as *Acervularia*. Corals consist of individual animals called polyps that produce a hard outer cup-like casing called a corallite. Fossil corals consist of either individual corallites or colonies of tightly packed corallites. Corals were the dominant components of the early patch reefs, just as they are in modern reefs, but the corals dating from the Silurian are different from living ones. Tabulate and rugose corals all became extinct 250 million years ago and were replaced by scleractinian corals. Although modern and extinct corals can look physically similar,

surprisingly they are only distantly related. Other common corals found in Wenlock Limestone include the tabulate colonial coral *Heliolites*, and the very distinctive tabulate coral *Halysites*, commonly known as chain-coral.

The Wren's Nest National Nature Reserve in Dudley is another famous outcrop of Wenlock Limestone and a great place for fossil-hunting. More than 700 different species of fossil plants and animals that once lived in the Silurian seas have been found here. In 1956, it was the first National Nature Reserve to be chosen on the basis of its geology, effectively protecting an important area of the limestone from further expansion of the housing estates that surround it on all sides. At first glance, the landscape at Wren's Nest is rather

Desmidocrinus macrodactylus *is one of the many species of fossil crinoid found in Wenlock Limestone. Commonly known as sea lilies, crinoids are, in fact, animals that flourished in the Silurian seas.*

confusing due to a long history of quarrying, first for building stone, then agricultural lime, and up until 1924, as a flux for the Black Country iron industry. What remains is a scarred landscape of trenches, tunnels, hillocks and tailing piles left behind by the miners. There is even an underground canal that was used to transport the quarried limestone to the kilns.

Any fossil enthusiast is welcome to rake through the loose gravel at the base of the slopes in the Nature Reserve; common finds include beautifully preserved brachiopods. Many different species of these small, shelled creatures thrived in the shelter of the patch reefs. You are also likely to come across small discs, like tiny fossilized Polo mints, that once made up the stems of crinoids. Crinoids, when complete, are some of the most beautiful fossils found in Wenlock Limestone. They are called sea lilies because they can look a bit

RIGHT: A reconstruction shows a trilobite scavenging for food on a sandy sea floor. Trilobites had rows of short legs on the underside of their bodies, but it is usually just the hard exoskeleton that is fossilized.

FAR RIGHT: The famous fossil trilobite Calymene blumenbachii *was nicknamed the Dudley Bug or Dudley Locust because it was so commonly found at Wren's Nest. The name trilobite, meaning 'three-lobed', refers to the way the body (thorax) is divided lengthways into three segments.*

like flowers: many had a long flexible stem rooted to the sea bed with a small cup at the top and many feathery arms. However, crinoids are actually animals that used their arms to waft food into a mouth inside the cup. They lived in huge numbers around the patch reef in 'crinoid gardens'. There are some living crinoids, but far fewer than there were in the Silurian period. Particular layers of Wenlock Limestone are made up almost exclusively of the remains of crinoids such as *Marsupiocrinus* and *Grissocrinus*. In the nineteenth century, quarrymen at Wren's Nest supplemented their income by selling complete crinoids and other attractive fossils to collectors and museums.

Wren's Nest is most famous for fossil trilobites, particularly *Calymene blumenbachii*. This trilobite was so commonly found by miners in the nineteenth century that they nicknamed it the Dudley Bug or Dudley Locust and it was incorporated in the town's coat-of-arms. Trilobites are arthropods, relatives of modern crabs and lobsters, but without any claws.

Many different species lived in the Silurian seas, crawling about on the sea bed scavenging for food, and they were typically a few centimetres long. As they grew, they periodically shed their hard outer skeleton, so you are more likely to find a moulted fragment than a complete specimen. Some species could roll up, just as some woodlice do today, and can be found fossilized in this position. Finding a Dudley Bug might not be as easy as it once was (no hammering is permitted at Wren's Nest without special permission), but the Wenlock Limestone is still one of the best rock types in Britain for fossil-detecting and provides some of the best evidence of ancient reef communities anywhere in the world.

A typical ammonite fossil from the Wiltshire dig site. The shell is crushed but retains a beautiful pearly, iridescent sheen.

An overview of the Wiltshire dig site shows the team from the British Geological Survey hard at work prising open slabs of Oxford Clay in the hope of finding fossils with exceptional soft-tissue preservation. Any significant finds are recorded and conserved at the bench, ready to be taken back to the laboratory for further analysis.

Wiltshire's Lost Fossils

RAILWAY DISCOVERIES

The rapid expansion of the railway and canal network in the nineteenth century, and all the building and quarrying that went with it, meant that many new fossil-hunting sites were opened up. Fresh and exciting discoveries from across Britain fed a growing appetite among both scientists and the public for further insights into the past and the strange creatures that lived in prehistoric times. The extension of the Great Western Railway into Wiltshire in the 1840s led to one such discovery that has become legendary in palaeontological circles. Exceptionally well-preserved fish and squid-like creatures were found next to a railway embankment being built near Chippenham. The site became famous for the stunning fossils it produced. It was equally notorious because after just a few years it was abandoned, somehow forgotten and lost. Around 150 years later, a team from the British Geological Survey (BGS) led by Dr Phil Wilby decided the time had come to renew the search for the lost Wiltshire site and to bring our understanding of its amazing fossils up to date.

The rocks in the Chippenham area date from the mid-Jurassic, about 160 million years ago. At this time, much of central England was submerged beneath the sea, and layers of mud and clay gradually accumulated on the sea floor. These sediments eventually formed Oxford Clay, a widespread sedimentary rock that can be found right across Britain in a broad stripe from Dorset to Yorkshire. Oxford Clay is full of fossils and is particularly rich in bivalves, ammonites and belemnites, as well as fish and large marine reptiles such as ichthyosaurs.

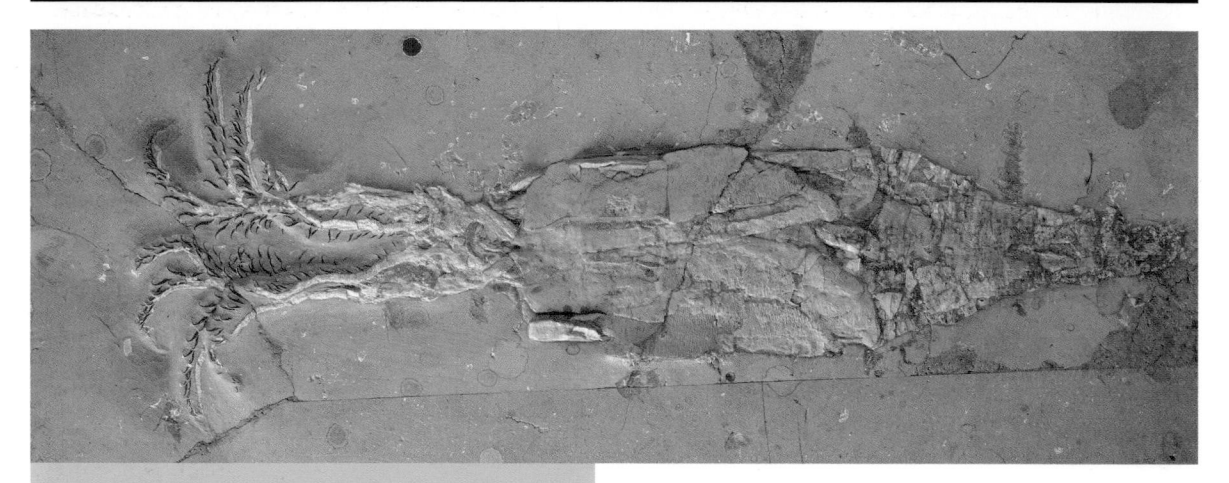

TOP: A model reveals what Belemnotheutis antiquus would have looked like when alive. It is grasping a fish in one of its arms.

ABOVE: Belemnotheutis antiquus is one of the exceptional squid-like fossils found in the mid-nineteenth century, with soft-tissue parts, including all the arms with hooks, preserved.

Under normal conditions, only the hard parts of these creatures were fossilized but, in that particular layer of Oxford Clay found near the railway bank in Victorian times, the soft parts of some animals were preserved in astonishing detail.

Among the remarkable fossils originally found were more than 100 coleoid cephalopods, which are related to modern squid. These included stunning specimens of *Belemnotheutis antiquus,* in which all the soft tissues such as the muscles, ink sac, arms, suckers and hooks had been preserved. This style of preservation is very rare and has been nicknamed the 'Medusa effect' because so much of the animal has been literally turned to stone in meticulous detail. During fossilization, the soft tissues of the dead animals were replaced by the phosphate mineral apatite. The exact process is not yet fully understood, but it seems that in this case bacteria were involved: excess phosphate in the water surrounding the animal or in the carcass itself stimulated bacteria to precipitate apatite, which hardened the soft body parts before they decayed. Understanding this unusual style of preservation in more detail is one of the aims of Dr Wilby's and his colleagues' research.

Many of the original nineteenth-century specimens ended up at the Royal College of Surgeons in London, and were destroyed in bombing raids during the Second World War.

Some still exist in museums, but they are so rare and important that no one would want to risk damaging them by subjecting them to modern investigative techniques. Even if Dr Wilby could use the original finds, Victorian scientists 'improved' the appearance of them by scraping away any other fossil evidence that might have existed on the surrounding slab of clay and then coating the specimens in a resin or varnish that cannot be removed. So although they have been invaluable for examining the biology of the animals, to fully understand them and place them in context, the only option for Dr Wilby and his team was to track down the original site. After sifting through historical records and working out which layers of Oxford Clay the fossil coleoids were most likely to have come from, the team drilled boreholes in various fields near the railway in Wiltshire in an attempt to locate the famous site. They were pretty sure that the relevant clay layer would be marked by a particular species of ammonite, and not only did they find the right layer, but one borehole cut through a fossil coleoid.

The exact location of the field site remains a closely guarded secret due to the enormous historical and scientific importance of the fossils. Many tonnes of Oxford Clay were excavated from a massive hole in the ground and arranged in piles according to the depth the clay had been dug from. Almost every slab contained fossil shells of bivalves, ammonites and spiky-shelled marine snails. Dr Wilby and his team prised apart the slabs of clay in the hope of finding some examples of soft-tissue preservation. Surprisingly, finding beautiful specimens such as those discovered by the Victorians was not their top priority: in the short time available, the team wanted to find as many soft-tissue samples from the widest range of animals possible. During the week-long dig, they uncovered over a dozen well-preserved shrimps complete with appendages/antennae, several fish specimens with preserved muscles, a complete ink sac and numerous nearly complete squid-like creatures. These samples were just what the team needed for their further studies.

After all the samples have been properly recorded and prepared at the BGS headquarters in Nottingham, the next stage of the project will be an analysis of the material by Dr Wilby's team as well as other specialists in Britain and abroad. The results will provide an insight into how conditions in the Jurassic seas at that particular place and point in time led to such incredible preservation. Chemical analysis that would have been impossible on the old varnished specimens will help to work out conditions in the ocean when the animals were still alive. The complete preservation of soft parts means that the fossils can be dissected like a modern animal. This will help reveal their anatomy, bodily functions and how they moved. Dr Wilby already knows that the preservation of muscle fibres in his new fish specimens is so good that under a scanning electron microscope it is hard to distinguish between a sample from a modern fish and a fossil fish that is more than 160 million years old. Now the site has been rediscovered by the BGS, it is unlikely that its location will be forgotten again for a very long time.

Dr Phil Wilby of the British Geological Survey scans fossilized fish muscle from the Wiltshire site under an electron microscope, revealing the astonishingly detailed preservation in the 160-million-year-old specimen.

Eve Roberts' Mammoth

AN ICE AGE MAMMOTH IN BRITAIN

One pleasant, late September Saturday afternoon in 1986, a Shropshire housewife, Eve Roberts, was taking her dog on one of their regular walks near the village of Condover. Their path took them past a local sand and gravel pit, which was usually bustling with heavy machinery during the working week.

As Eve cast her eye over the workings, she noticed a curious brown object sticking out of the muck. At first, she thought it was part of a rusting motorbike that had been dumped in the quarry. But on closer inspection she could see a big, curved sheet of bone. With her curiosity aroused, she contacted the local museum.

That evening, Eve showed the bone to Jeff McCabe of the Shropshire Museum Service. When they looked around they found others, including one that was 1.2 metres long and rather like a giant human thigh bone. The odd thing was that the bones looked incredibly fresh, unlike most fossilized bone material, which becomes mineralized and rock-like.

Knowing that this was indeed worth further investigation, McCabe contacted Dr Russell Coope, an academic at Birmingham University and a well-known expert on Ice Age life and environments. From McCabe's description, Coope knew that this was probably a big mammal and enlisted the help of Adrian Lister, a young researcher at Cambridge University who was fast making a name for himself for his work on the big beasts of the Ice Age, the so-called 'megafauna'.

Since the Condover sand and gravel quarry was actively being worked, McCabe, Coope and Lister knew that they had no time to waste. They visited the site and confirmed that the bones belonged to an adult mammoth and were coming from a muddy pile that had recently been bulldozed to one side. Evidently, the original skeleton lay somewhere else in the quarry. The only chance of finding the remainder was to gather all the mud so that it could be systematically searched.

The experts rustled up an army of willing volunteers to help them search methodically through the sticky clay. Although progress was slow, a number of bones of different shapes and sizes emerged. To everyone's delight, a small but entire jawbone appeared out of the muck. It soon became obvious that there was more than one mammoth – a juvenile as well as an adult.

Eventually, around 400 bones of all shapes and sizes were recovered. When finally reassembled, it turned out

Just over 12,000 years ago the mammoths from which this lower jaw and vertebra came were alive and well in post-glacial Shropshire.

that most of the adult skeleton had been preserved (apart from the skull and small foot bones) along with bits of at least three baby mammoths. From examination of their teeth, Lister estimated that they were aged between three and six years old, while the adult was a mature bull male around 26 years old. For fossil detectives the big question is, of course, how and why did these mammoths die? Was it of natural causes or was something more sinister at work?

Examination of the quarry site showed that it had originally contained what is known as a 'kettle hole'. These are common features of glaciated lowland landscapes plastered over with glacial debris. Great blocks of ice were often caught up among the debris and, when they melted, they formed large, deep holes that filled up with water. In this case the hole was about 10 metres deep at its centre and was surrounded with succulent marsh plants, which probably attracted the plant-eating mammoths.

Although mammoths could swim, the steep, muddy sides of the kettle hole would have been difficult, if not impossible, to climb out of. Perhaps the male was attracted by the squeals of the foundering babies who had slipped in, and himself fell while trying to help them, although there is no proof that the drowning of the babies and adult were connected.

The heavy head and tusks probably became detached from the rest of the body, and have never been found. But the cold water helped preserve the rest of the skeleton in almost pristine condition. Radiocarbon dating of the bones showed that the animals lived between 12,700 and 12,300 years ago – just after the last glaciation – making them the most recent mammoths ever found in Britain. So it appears that British mammoths managed to survive that last cold glacial event, but not for long. For some reason they were not able to re-establish themselves as a viable population.

The woolly mammoth Mammuthus primigenius *was well adapted for life on the margins of Ice Age Britain. Small ears and tail reduced the risk of frostbite, the long body hair was an excellent insulation against the cold, and the curious-looking 'humps' were fat stores.*

WALES

In 1835 a geologist called Adam Sedgwick defined a new division of geological time based on his studies of the rocks and fossils in northwest Wales. He called it the Cambrian after the Roman name for Wales, Cambria. At the same time, Roderick Murchison defined the Silurian period based on the rocks and fossils in southern and eastern Wales, naming it after the Silures, a Welsh hill tribe. After initial agreement, a dispute about where the Cambrian ended and the Silurian began divided the two geologists. During his lifetime, Murchison drummed up considerable support for his scheme of an extended Silurian period. In the end, after 40 years of controversy, Charles Lapworth proposed a new period called the Ordovician (named after the Ordovices, another Welsh tribe) between the Cambrian and Silurian, effectively solving the problem. The divisions of all three periods have been refined over the years, and they still make up the Lower Palaeozoic era, from 542 to 416 million years ago.

The Cambrian marked a very significant phase in the evolution of life, when there was a rapid diversification of complex life forms, especially those with hard skeletons. The so-called

Southerndown Beach at Dunraven Bay on the South Wales coast is a good place to find Jurassic fossils such as the oyster Gryphaea arcuata, known as 'Devil's toenails'. Keep clear of the cliffs in case of rock falls.

Cambrian 'explosion of life' is well documented in Welsh rocks. Many of the fossils used by Sedgwick and Murchison to illustrate their ideas were trilobites. One of the largest trilobite species can be found on the south coast of Wales. *Paradoxides*, from the Cambrian, grew to 50 cm long. Other fossil finds are the basis for stories in this chapter, including Triassic dinosaur tracks, rare Jurassic molluscs and a much more recent human skeleton.

Silica Fossils

THE CASE OF THE 'MISSING MOLLUSCS' UNDER THE FORD FACTORY AT BRIDGEND

The Ford engine factory at Bridgend in South Wales produces almost a million car engines a year – all, of course, destined to run on fossil fuels. But that isn't the reason for mentioning them here. The real story began when the foundations for the engine plant were dug in the 1970s and the builders unearthed a very rare assortment of fossils. The only fossil left at the factory site today is a single large ammonite proudly displayed in the company's boardroom. All the others, mainly of smaller sea creatures such as bivalves and brachiopods, are in the care of scientists at Cardiff University and the National Museum of Wales.

The fossilized animals all lived in an ancient, deep sea that covered large parts of Britain in the Jurassic period, about 170 million years ago. When they died, the animals sank to the bottom and were buried in the sea-floor sediments that eventually became the rocks in parts of South Wales. There is nothing particularly remarkable about that, one might think, as rocks of this type and age commonly yield fossil marine animals.

Rare silicified fossils were found on the site of Ford's Bridgend Engine Plant when building work began in the 1970s. The photograph shows part of the assembly line.

RIGHT: *Silicified ammonites from Bridgend delicately preserved in three dimensions.*

BELOW: *A cluster of silicified Gryphaea and other bivalves found beneath the Ford factory. The fossils were separated from the surrounding sediment by putting them in acid, which dissolves the sediment.*

What makes these fossils special is that they had been silicified. This means that the mineral silica had replaced parts of the animals, particularly their shells, during fossilization. It is an unusual form of preservation, and silicified fossil assemblages can include many species that are normally missing from the fossil record. Consequently, collections like the one found under the Bridgend Ford factory are highly sought after.

When scientists at Cardiff University compared the Ford fossils with another typical, but unsilicified, Jurassic sea-floor assemblage, they found something remarkable. Many fossil animals that are usually absent elsewhere, including whole groups of burrowing bivalves, are preserved at Bridgend. In fact, without the silicification, it is estimated that 65 per cent of the bivalve mollusc fossils would be missing. These 'missing molluscs' can be studied alongside all their relatives in the Bridgend collection, providing palaeontologists with a much truer picture of life in a deep Jurassic sea. It seems that life there was more diverse than previously imagined.

But why are many shelled fossil creatures often missing in other places? The problem is that marine creatures have shells that are made of calcium carbonate that comes in two forms: the mineral calcite or the mineral aragonite. Aragonite often dissolves in tens to hundreds of years when buried in sea-floor sediments, particularly in deep water. This is much quicker than shells made of calcite, so fossil assemblages of marine invertebrates tend to contain a disproportionate number of calcite-shelled species compared to aragonitic ones. But the fossils from Ford are different: when the shells of the dead animals sank into the muddy sea floor all those millions of years ago, water rich in silica percolated through the sediments. Silica seeped into the shells and replaced the original material of both the calcite and aragonite-shelled species, preserving them all.

The silica may have leached out of older Triassic rocks underlying the Jurassic sediments, or given that preservation must have taken place quickly before there was time for the aragonite to dissolve, perhaps there were hot springs on the sea bed that provided the silica-rich water.

Palaeontologists have to work hard to get a true and undistorted picture of the diversity of life that existed in the past. Bias in the fossil record towards those creatures that have better 'preservation potential' than others is a major problem. Those with hard body parts are more likely to become fossilized than soft-bodied creatures, but even though shelled animals are good contenders for fossilization overall, within these animal groups there is still a problem of uneven representation. The fossils from the Ford factory show us that animals with aragonite shells are likely to have been more prevalent in the past than we routinely see elsewhere.

Some of these 'missing molluscs' are being studied at the National Museum of Wales. Apart from being scientifically important, they are also quite beautiful. The silica has preserved the shells in delicate 3D; intricate surface details are visible and even some soft internal body structures such as

feeding tubes are preserved. The fossils have to be handled with tweezers and great care taken not to break them. Some of the shells are so fine that when held against a light they look like gold-dipped lace. These fragile fossils offer a rare insight into the diversity of life forms in Jurassic seas and, for the scientists that study them, they are a chance to overcome bias in the fossil record. They are also a reminder that the most important fossils can be found in the most unlikely places.

An artist's impression of a scene from the Jurassic seas showing an empty ammonite shell sinking to the bottom under the watchful gaze of a plesiosaur.

Bendrick Rock

FOSSILIZED DINOSAUR FOOTPRINTS

The best place in Wales to see fossilized dinosaur footprints is Bendrick Rock near Barry on the south Glamorgan coast. Once you start looking it's not hard to spot dozens of conspicuous three-toed impressions on the red rocky ledges that flank the beach. If the prints are full of water, they create exquisite dinosaur-footprint puddles. The site on the Welsh coast is a treat for keen fossil-hunters, as the fossils are freely accessible for anyone to go and have a look. It is thrilling to walk, quite literally, in the footsteps of dinosaurs that once roamed this part of South Wales during the late Triassic period, about 220 million years ago — a humbling experience that certainly puts the brief human period into perspective.

Bendrick Rock gained notoriety in 2006 when as many as 20 of the dinosaur tracks were stolen. Thieves had used rock saws and crow bars to illegally remove slabs of rock containing the precious prints. The Countryside Council for Wales was tipped off when some of the rocks ended up for sale on the internet. Fortunately, all the fossils were recovered, and the unscrupulous collector was cautioned by the police. The slabs cannot be put back, but at least now they are safe and will be housed in a museum for all to enjoy. There might be slightly fewer left to see in situ, but Bendrick Rock is still rich in these fossils and definitely worth a visit.

In Triassic times, much of Britain was a hot, dry desert near the middle of an enormous supercontinent called Pangaea, and much closer to the equator. The climate was arid, but occasional rain storms meant that the sandy landscape was inundated by flash floods. At the first signs of water, dinosaurs congregated around ephemeral streams and pools to quench their thirst and left their footprints in the wet sand. As the water evaporated, the prints were baked hard by the sun and then covered by sediments washed in by the next flood or blown over them by the wind. The impressions have been preserved between layers of sand that eventually formed the red sandstones exposed along the South Wales coast. Erosion by the sea has spilt apart the rocks, revealing the tracks at the surface once again.

Dr Phil Manning from the University of Manchester — a regular dinosaur expert on *Fossil Detectives* and an authority on interpreting dinosaur trackways — explains that the majority of prints were made by bipedal predatory dinosaurs such as *Coelophysis*.

Layered Triassic sandstone at Bendrick Rock on the south Glamorgan Coast. The site is freely accessible and the best place in Wales to see fossilized dinosaur tracks, but beware – the rock ledges can be very slippery.

Some of the very clear specimens are about 12 cm long but there are also larger tracks, possibly made by a prosauropod dinosaur, similar to *Plateosaurus*. As no bones have ever been found at this site it is hard to be absolutely sure. They are the oldest-known dinosaur tracks in Britain and were made by some of the first dinosaurs to evolve, near the beginning of their reign as the most important animals of the Mesozoic era.

Fossilized tracks, such as those found at Bendrick Rock, can yield astonishingly detailed information about the animals that made them. For example, a dinosaur's gait and posture can be worked out from its tracks. Recent headline-grabbing research by Phil and his co-workers has taken this process

One of several fossilized dinosaur footprints visible at low tide. This one is filled with water and about 12 cm long. It was made by Coelophysis, a bipedal predatory dinosaur that lived in the area about 220 million years ago.

one step further: using a technology called evolutionary robotics, they have worked out exactly how bipedal dinosaurs moved and how fast they could run. This is an important element in understanding their behaviour; predator-prey relationships, for example, depend on relative running speeds.

The University of Manchester team has developed a computer model of the skeletal and muscular structure of five different dinosaurs based on information gleaned from fossil bones and tracks from many different sites around the world. The model was used to map out the optimal biomechanics of each animal to help define their top running speeds. It was a very long and complex task, even for a supercomputer, but the results were to reveal some startling facts about the movement of these ancient creatures.

The model included many variables, such as the forces generated by muscles, the spring recoil of tendons and the constraints imposed by limb joints. To check the model was working, the results for the five extinct dinosaur species were compared with model results for extant (living) bipedal species including humans. The scientists effectively ran a virtual race to see who won.

The model shows that, as expected, the bigger the dinosaur, the slower it ran – but even the mighty *Tyrannosaurus rex* (*T. rex*), the biggest of the animals modelled weighing in at 6000 kg, could have chased down all but the fastest humans. In the model, *T. rex* reached speeds of almost 18 mph. That is slower than an Olympic sprinter, who can reach speeds of around 22 mph, but marginally faster than most extremely fit individuals, such as professional football players. Despite their size, these creatures could certainly get around.

The smallest dinosaur in the model, *Compsognathus*, a turkey-sized creature that lived in the Jurassic, reached an exceptional top speed of almost 40 mph. That is faster than any living two-legged animal, including ostriches and emus. Other species, such as *Velociraptor* (which weighed 20 kg) and *Allosaurus* (1500 kg), could run up to 22 and 25 mph respectively.

The team at the University of Manchester has revealed new information about the athleticism of dinosaurs. No living animals move in quite the same fashion, so using the fossil evidence in combination with computer modelling is the only way to investigate the mobility of these ancient creatures. It seems that many dinosaurs, especially the predators, were fast runners – not the lumbering, dim-witted beasts they were thought to be when first discovered. It is a staggering feat for creatures of this size and would make alarming viewing. Fortunately, today we can marvel at the Bendrick Rock fossils safe in the knowledge that none of us will ever have to outrun a dinosaur for real.

An impression of what Coelophysis, one of the dinosaurs that left their tracks at Bendrick Rock, may have looked like. Coelophysis is one of the earliest known dinosaurs and may have hunted in packs. It grew up to 3 metres long and stood about 1 metre tall at the hip.

Fossils and Folklore

ANCIENT TALES ABOUT FOSSIL LIFE

It was not until the mid-eighteenth century that our current understanding of fossils as a record of ancient life started to be widely accepted by academics and then more gradually by the general public. Before that time, people had come up with a whole host of theories and beliefs in an attempt to explain what fossils were and what they might be useful for. From natural curiosities to divine creations, lucky charms and cures for ailments, fossils have been thought of in a variety of ways by our ancestors.

If you travel around Britain you may hear a few traditional stories associated with fossils. Colloquial names such as 'thunderbolts' and 'devil's toenails' hint at old myths and legends. One person who is well versed in such tales is Professor Mike Bassett, Keeper of Geology at the National Museum of Wales, and President of the Palaeontological Association.

Mike has many stories about ammonites, surely the most recognizable of all fossils. Their coiled shells are an enduring fossil icon and adorn many things, from the lamp posts in Lyme Regis to countless books and websites. As Mike explains, their name originates from ancient Greece, where they were thought to resemble a ram's horn, which

Professor Mike Bassett and Hermione at Caerphilly Castle. On the bench between them is a large example of a carved snakestone.

The tradition of carving snake heads onto ammonite fossils, as seen here on a specimen of Dactylioceras commune, *may have started in Whitby on the North Yorkshire coast, where this one comes from.*

was also the sacred symbol of the ancient god Ammon. Specimens were sometimes known as *Cornu Ammonis* (meaning 'horns of Ammon') before their current scientific name became common.

We now know that ammonites are extinct marine molluscs that thrived during the Mesozoic era, but throughout British history they were widely known as 'snakestones'. Many people believed that they were petrified snakes and although their heads were never found, carved heads and tails were crafted onto specimens.

This tradition is strongly linked to a legend from Whitby in Yorkshire. The story goes that lots of snakes once inhabited the area, but the seventh-century Saxon Abbess St Hilda turned them to stone as she cleared them from the land she wanted to build her convent on. The snakestone became an emblem of the town sometime during the sixteenth and seventeenth centuries and was used to decorate trading tokens and the town's coat of arms.

There are examples of carved snakestones from all over Britain, including many of the ammonite species *Hildoceras bifrons*, which is named after St Hilda.

Thunderbolts was the popular name for the pointed, cigar-shaped fossils of belemnites that are very common in Jurassic- and Cretaceous-aged rocks in Britain. They were thought to have been cast down from the heavens during thunderstorms, and their name is derived from the Greek word *belemnon,* meaning 'dart'. They are actually the fossilized

Cretaceous-aged belemnite fossils from Britain, popularly known as 'thunderbolts', are the fossilized internal guards of extinct squid-like creatures. They were thought to have a range of medicinal powers.

internal guards from an extinct group of cephalopods related to modern-day squid and cuttlefish. In various parts of Britain they were also known as 'Devil's fingers', 'St Peter's fingers' or, more logically, as 'bullets'.

As Mike explains, belemnites – more than any other group of fossils – have been associated with medicinal powers. In southern England they were used as a cure for sore eyes and blindness. The treatment involved crushing the fossils and blowing the dust into the eyes, not a practice that bears up to the scrutiny of modern medicine. In Scotland, belemnites were called 'bat stones'. They were thought to cure worms and distemper in horses if the animals drank water that belemnites had been soaked in. In other places, crushed belemnites were thought to cure cow's colic. Perhaps there's an element of truth behind this claim because belemnites are made of calcium carbonate (chemical formula $CaCO_3$), which is a well-known cure for indigestion.

Echinoids, or sea urchins, are another type of fossil commonly found in Britain that has great historical

Echinoids, or fossilized sea urchins (like this Echinocorys from southern England), were popularly known as 'fairy loaves'. In the past, people often kept them in their houses as a lucky charm to help ensure successful baking.

significance, and many different folklore stories are attached to them. Almost 100 were found arranged in circles around the bodies of a woman and child in a 3000-year-old grave on Dunstable Down, near London. It is probable that echinoids were considered useful or valuable in the afterlife. In Medieval times, they were known as 'snakes' eggs'. One version of this myth has it that echinoids were formed from froth formed by snakes congregating at midsummer. The eggs were thought to retain magical powers if they were stolen from the snakes, taken across a river where the angry snakes couldn't follow and kept on a piece of cloth.

In more recent times, echinoids with a distinctive domed shape (particularly species of *Conulus* common in the Chalk of southern England), were called 'shepherds' crowns' or 'pixies' helmets'. It is likely that shepherds would have come across them weathering out of the underlying Chalk. In some places, especially Suffolk and Sussex, they were known as 'fairy loaves'. Stories suggest that loaf-shaped echinoids such as *Echinocorys* or *Micraster* were kept in people's houses to ensure successful baking or to stop milk going sour, and to help protect the family against witchcraft.

Mike has many other fossil folklore stories to share. In the Carmarthen district of South Wales for example, the tails of some Ordovician trilobites were once thought of as butterflies that had been petrified by the magician Merlin. The common fossil oyster *Gryphaea arcuata* was known as a 'Devil's toenail', presumably because it has a curved shape and layered structure reminiscent of a nasty, curling, flaky toenail. In Scotland, *Gryphaea* are called 'crouching shells' in Gaelic, and were once thought to cure arthritis.

Folklore stories illustrate how much our perception of fossils has changed over the years. Modern palaeontological discoveries mean that they are no longer mere curiosities or mythical objects. Their place in the story of the evolution of life gives them far more significance than ever before. But fossils can still be appreciated as objects of beauty and considered as natural treasures. Fossils connect us to lost worlds that we could never experience otherwise. In a way, they allow us to travel through time – so perhaps they do have some magical qualities after all.

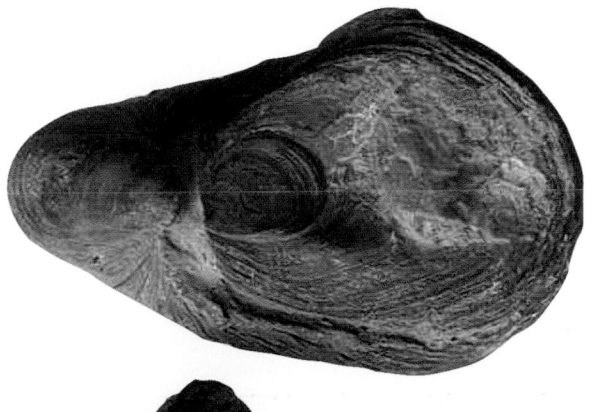

The curved fossilized shells of Gryphaea arcuata, *an oyster that lived on the muddy sea bed in shallow seas during the Jurassic and Cretaceous periods. They were popularly known as 'Devil's toenails', and were once thought to cure arthritis.*

The Red Lady
OF PAVILAND CAVE

The discovery in the 1820s of a human burial in a cave on the South Wales coast was potentially one of the most important breakthroughs in the story of human settlement in Europe. But its scientific importance was not realized at the time. The initial interpretation of this find by the Reverend Doctor William Buckland – one of the heroes of nineteenth-century British palaeontology – reflects beliefs that were commonly held in his time, which were that fossilized human remains could never be found.

Goat's Hole Cave is now on the coast, but 26,000 years ago the limestone cave provided an excellent vantage point for human 'big-game' hunters overlooking a wide coastal plain.

In the 1820s, Buckland was particularly interested in cave deposits and how they might provide good fossil evidence as proof of the Noachian Flood as described in the Old Testament. He was one of the so-called 'theological geologists' who hoped that the science of geology would vindicate the biblical version of Creation and the Flood.

Buckland had just excavated Kirkdale Cave in Yorkshire, where he had found over 300 hyena teeth along with the remains of some 20 species of mammal, including hyenas, rhinoceroses, bears and big cats, among others (see pp. 82–3). He thought these animals had lived near the cave just before the Flood, around 6000 years ago. Buckland associated tooth marks on the bones with the bone-chewing habits of hyenas and suspected they had dragged the bones into the cave to consume at their leisure. His theory supposed that the Flood had then swept all evidence away from around the cave but had not reached the bones buried deep inside. Buckland's brilliant analysis of the remains and their behavioural interpretation was one of the first ecological analyses in the history of palaeontology and earned him the Copley gold medal of the Royal Society.

Paviland Cave, near Swansea in South Wales, also contained many animal bones, including those of extinct animals, but here they were accompanied by artefacts made from bone and ivory, stone tools and an almost complete human skeleton. Buckland carried out what was an unusually careful excavation for the time and recovered more than 5000 artefacts. He described the human skeleton as that of a modern-looking, slenderly built human female, 1.7 metres tall. He nicknamed her the Red Lady of Paviland because the body was found in a shallow grave, covered with a dusting of red ochre. The grave also contained long, thin wands, bracelets and a pendant made of mammoth ivory, along with seashells perforated as if for a necklace. Buckland could see that the body had been buried with considerable ceremony but he was puzzled by the association of the skeleton with extinct animals such as the mammoth.

To explain the situation, Buckland concocted a convoluted story – the Red Lady was a young member of a Welsh Celtic tribe that lived in the area during the Roman occupation. How she died was not known but for some reason her kinsfolk buried her discreetly in the cave, where they found some mammoth ivory from an animal that had died in the Flood. They then proceeded to carve objects from the ivory and place them in the grave.

Buckland was convinced that humans were specially created after all the other animals and agreed with the French anatomist Georges Cuvier that, as a result, no fossil human remains would ever be found. Even when he was confronted with the blatant evidence that humans had existed alongside extinct animals, such as mammoths, he could not take in what he was seeing.

We now know that the Red Lady was in fact a young man, about 25 years old, who lived some 26,350 years ago, just before the last glaciation. He was one of the Cro-Magnon people who first

Buckland's excavation of the cave burial was one of the first to be carried out in a reasonably methodical and well-recorded scientific way.

migrated into Europe some 38,000 years ago. When this man was alive, sea levels were lower than they are now and Paviland Cave was separated from the Bristol Channel by a wide, marshy coastal plain. Herds of animals used the plain as a source of food and a migratory route, so the cave was well situated for the Cro-Magnon hunters to watch and select their prey.

Buckland and his contemporaries may have been astounded to find human bones next to the remains of ancient creatures, but this tale is a tribute to our ancestors, who really did walk on Earth all those thousands of years ago.

The leg bone of Buckland's Red Lady of Paviland turned out to be that of a young man, and no different anatomically from those of modern humans.

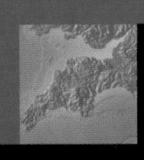

Southwest ENGLAND

The southwest of England is world famous for its outstanding geological heritage. In 2001, 95 miles of Dorset and east Devon coastline officially became England's first natural World Heritage Site, joining the ranks of St Kilda in Scotland and the Giant's Causeway in Ireland as an internationally recognized natural wonder of the world.

The site is popularly known as the Jurassic Coast, but it might be more accurately called the Mesozoic Coast: between the western extreme near Exmouth and the eastern end at Purbeck, the rocks represent 185 million years of Earth history, spanning the older Triassic and younger Cretaceous periods too.

The Jurassic Coast is arguably home to the best places in Britain for finding fossils. At Lyme Regis, the dark cliffs are formed from Lower Jurassic marine rocks and have been a constant source of spectacular fossil finds for centuries. Many important discoveries were made by a remarkable woman called Mary Anning in the early nineteenth century. Despite years of collecting, high rates of erosion and cliff collapse mean that new discoveries are still being made. If you visit this spectacularly beautiful

Fossil-hunters take advantage of a cliff collapse at Charmouth on the Dorset coast to look for newly exposed specimens. Over the years, many spectacular fossils have been found in these unstable cliffs.

Mary Anning

THE PRINCESS OF PALAEONTOLOGY

One of the most successful fossil-collectors of all time was born in Lyme Regis in 1799. Her name was Mary Anning. Mary made a meagre living from selling fossils, but despite her poor background she knew as much about the anatomy of her specimens as the leading gentleman scientists of the day. Her impressive list of discoveries includes the first ichthyosaur, the first plesiosaur and the first British pterosaur known to science. As one of very few women to feature in the history of the subject, she has been called the 'Princess of Palaeontology'.

In the early nineteenth century, the small town of Lyme Regis on the Dorset coast became a fashionable seaside resort. Well-to-do tourists, including the likes of Jane Austen, would often buy fossils, or 'curiosities' as they were then known, as souvenirs of their visit. Palaeontology was just beginning to emerge as a new science, and wealthy gentlemen and university scholars were always interested in acquiring new fossils. Mary was born into an impoverished family that already collected fossils to supplement its income. Richard, her father, was a cabinet maker who died in 1810, leaving his family in considerable debt. With her mother and her only surviving sibling, Joseph, Mary collected and sold fossils as part of the family business from a young age.

This portrait of Mary Anning dates from 1847. Mary was the inspiration for the well-known tongue twister 'She sells sea shells on the sea shore'.

coastline, remember that the cliffs can be hazardous – it is safer and often more productive to stick to the beaches for fossil-hunting.

Such is the draw and wealth of fossil stories along the Jurassic Coast that you would be forgiven for thinking that there is nowhere else to go in the southwest, but in Somerset and Avon, particularly, there are sites well worth visiting, notably Writhlington in Somerset.

De la Beche's 1830 Duria Antiquior (A More Ancient Dorsetshire), from which this drawing is taken was the first attempt to depict prehistory from the fossil record.

The cliffs and beaches near Lyme Regis where Mary made her ground-breaking discoveries are still among the best fossil-hunting grounds in Britain. The rocks to the west of the town are limestones and shales of the Blue Lias group, and to the east towards Charmouth are the Black Ven marls. The rocks date from the early Jurassic period, about 180 to 200 million years ago, and are full of fossilized marine creatures from the Jurassic seas. The most common are ammonites and belemnites, but it was the discovery of large marine reptiles or 'seadragons' that made the Anning family famous.

In 1811, Joseph Anning spotted the well-preserved skull of a seadragon, and Mary helped him to collect it. By the following year, at just 12 years old, Mary had found the rest of the skeleton. The specimen, which was given its formal name of *Ichthyosaurus* a few years later, was sold for £23 to the lord of the local manor, Henry Henley. After changing hands several times, it now has pride of place in the Natural History Museum in London. This first ichthyosaur is often remembered as Mary's most significant discovery, but in fact she later found better ichthyosaur specimens (she sold a particularly good one for £100 in 1819), and more important discoveries were to follow.

One of Mary's best finds was a complete plesiosaur skeleton in 1823. Two years earlier, prominent scientists William Conybeare (1787–1857) and Henry De la Beche (1796–1855) had named and described the creature on the basis of a few bones that Mary had probably also found, but this new specimen caused a sensation. Some people, including the eminent French anatomist Georges Cuvier (1769–1832), thought at first it was a fake. It appeared to be an extraordinarily strange creature: almost 3 metres long with huge paddles, a long thin neck and a tiny head. Despite Mary's crucial role, no mention of who had found it was included in Conybeare's scientific account. Sadly, this happened time and time again, with Mary receiving very little credit for her work in any contemporary scientific literature.

As Mary grew up, she became the most important fossil-collector in the family. She learned to read and write, studied hard and corresponded with fossil enthusiasts all over the world. In 1828, she discovered the fossilized ink sac of an ancient squid-like creature called *Belemnosepia*. Amazingly, the fossilized ink could still be watered down and used, and apparently other similar discoveries led to sales of fossil-ink drawings in Lyme Regis shops. Mary also correctly

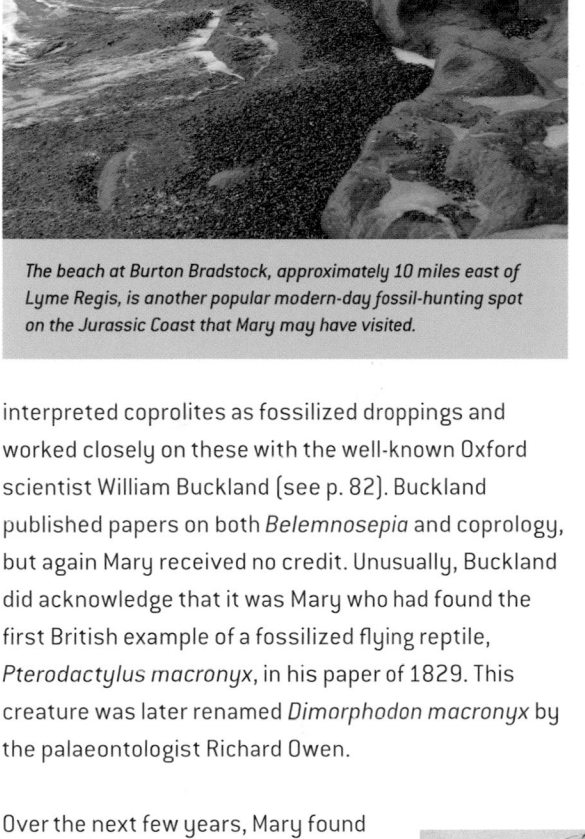

The beach at Burton Bradstock, approximately 10 miles east of Lyme Regis, is another popular modern-day fossil-hunting spot on the Jurassic Coast that Mary may have visited.

Many wealthy fossil enthusiasts visited Lyme Regis, as much to meet the famous Mary Anning as for anything else. It is through their letters and journals that science historians have pieced together many aspects of her life as reported here. Most visitors were surprised and impressed by how much Mary knew. Among her admirers was the King of Saxony, who bought a baby ichthyosaur from her for his collection in 1844.

In the final years of Mary's life, the scientific community clubbed together to support her, and some would say their generosity was well overdue. She received an annuity of £25 from funds raised at a meeting of the British Association for the Advancement of Science and a donation by the Prime Minister Lord Melbourne. Soon after her mother's death in 1842, Mary, who had never married, developed breast cancer and died in 1847, aged 48. Her long-time supporter Henry De la Beche published an unprecedented obituary in the *Quarterly Journal of the Geological Society*. Only fellows were supposed to receive this honour and no women were admitted to the society until 1904. Even then, no one as lowly as Mary would have been allowed to join. The full extent of Mary's contribution to palaeontology may never be known, but she is undoubtedly an inspirational and important scientific figure.

interpreted coprolites as fossilized droppings and worked closely on these with the well-known Oxford scientist William Buckland (see p. 82). Buckland published papers on both *Belemnosepia* and coprology, but again Mary received no credit. Unusually, Buckland did acknowledge that it was Mary who had found the first British example of a fossilized flying reptile, *Pterodactylus macronyx*, in his paper of 1829. This creature was later renamed *Dimorphodon macronyx* by the palaeontologist Richard Owen.

Over the next few years, Mary found other ichthyosaurs, plesiosaurs and a remarkable fossil fish called *Squaloraja*. Her last major discovery was in 1830: *Plesiosaurus macrocephalus*, a new species of plesiosaur. Mary had made quite a name for herself but, despite all this success, her earnings were unreliable at best. Among the people who helped her financially was Henry De la Beche. In 1830, all the proceeds from the first version of his celebrated illustration *Duria Antiquior* went to her.

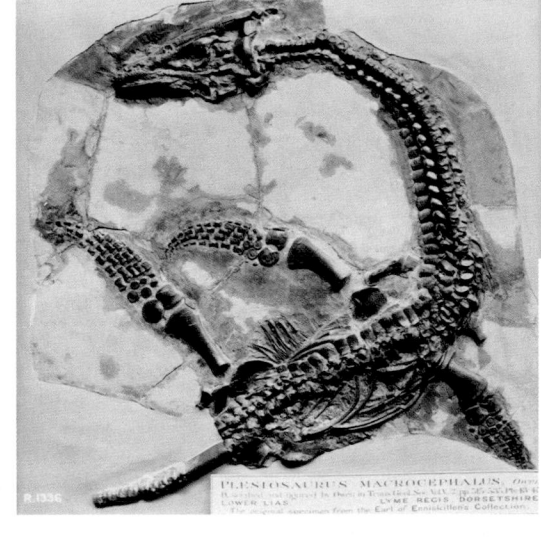

This large-headed plesiosaur Plesiosaurus macrocephalus *was found by Mary Anning in 1830 near Lyme Regis. The fossil is now in London's Natural History Museum.*

Scelidosaurus

THE CHARMOUTH DINOSAUR

Dinosaur fossils are highly sought after but are frustratingly scarce. This is because dinosaurs lived on land, where the right conditions for fossilization rarely occur. In contrast, the cliffs at Charmouth and many other locations along the Jurassic Coast are rich pickings for fossil-hunters because the rocks were created from sea-floor sediments. Any dead animal sinking to the bottom of the sea had a good chance of being quickly buried and was therefore more likely to become a fossil. It is relatively common to find swimming reptiles such as ichthyosaurs and plesiosaurs washed out of the cliffs at Charmouth, but very occasionally something even more exciting and rather unexpected turns up, such as a complete dinosaur.

In December 2000, an accomplished local collector, David Sole, was walking along the beach when a rock caught his attention. To the untrained eye it might have looked like an uninteresting, dull-grey lump, but David recognized its potential. Tapping it with his geological hammer revealed a piece of unmistakable dark-coloured bone trapped inside. Over the next few years, David and other collectors, including his son, searched the beach for more treasures and managed to piece together an astonishing fossil: a virtually complete skeleton of a dinosaur called *Scelidosaurus harrisonii*.

This is the best-preserved dinosaur ever found in Britain and it is a wonderful sight. The bones are now displayed at the Bristol City Museum and Art Gallery. There is also an excellent cast of the complete fossil

on the wall at the Charmouth Visitor Centre. The skeleton is about 2.5 metres long and at first glance looks rather complicated, with lots of lumps and bumps on top of the bones – small bony plates called scutes. These scutes were the dinosaur's body armour to protect it and ward off predators. Perhaps the plates could have functioned to attract a mate or to intimidate rivals. The scutes are similar to the smaller ridges on the back of a modern-day crocodile.

The scutes covered most of the body of the dinosaur and ran in rows from its head to its tail and down to its ankles. This specimen is unusual in that it has two horn-like scutes at the back of the skull that give

LEFT: *A selection of ammonites (each approx. 2 cm wide) and belemnite fragments collected from Charmouth beach. These are the most common fossils found on this stretch of coast.*

BELOW: *The fossil skull of the Scelidosaurus dinosaur found at Charmouth in 2000 is complete with serrated teeth and a large eye socket. Two horn-like scutes visible on the back of its head give the fossil its name: 'The Horned Scelidosaur'.*

the dinosaur its name: 'The Horned Scelidosaur'. These horns and the array of large scutes along its neck suggest that Sole's dinosaur was an adult male. *Scelidosaurus* lived about 195 million years ago. The bony armour plating puts it in a group of dinosaurs called Thyreophora, which includes the club-tailed dinosaurs known as ankylosaurs and the famous spiky-backed *Stegosaurus*, although the latter lived millions of years later.

Although these armoured dinosaurs looked ferocious, they were actually herbivores. *Scelidosaurus* had serrated teeth to cut up vegetation and fleshy cheeks to hold food in its mouth as it chewed. It probably had gastroliths, or grinding stones, in its stomach to help digest the plants it ate. Scientists may even be able to work out exactly the type of food this creature ate. It looks as though David Sole's fossil has the remains of its last meal still in its throat, suggesting it may have been sick just before it died.

With the help of local quarrymen (the unsung heroes of many fossil discoveries), Harrison found a few more bones of a young *Scelidosaurus* and eventually a nearly complete adult that is now displayed at the Natural History Museum in London. Years later, in 1985, another juvenile *Scelidosaurus* was found near Charmouth. Although incomplete, this discovery was exceptional because it included fossilized pieces of skin. This find is now housed at the Bristol City Museum and Art Gallery.

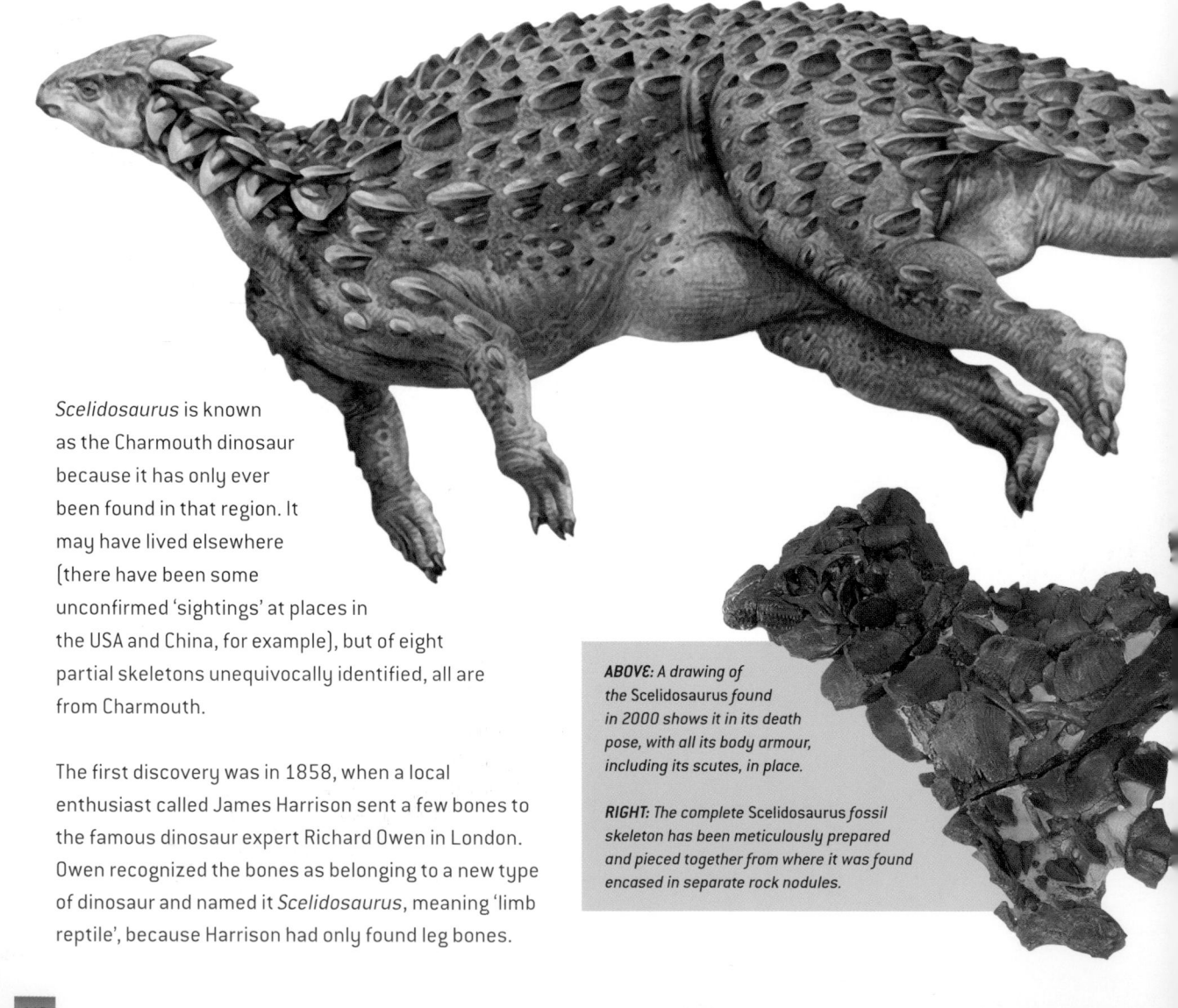

Scelidosaurus is known as the Charmouth dinosaur because it has only ever been found in that region. It may have lived elsewhere (there have been some unconfirmed 'sightings' at places in the USA and China, for example), but of eight partial skeletons unequivocally identified, all are from Charmouth.

The first discovery was in 1858, when a local enthusiast called James Harrison sent a few bones to the famous dinosaur expert Richard Owen in London. Owen recognized the bones as belonging to a new type of dinosaur and named it *Scelidosaurus*, meaning 'limb reptile', because Harrison had only found leg bones.

ABOVE: *A drawing of the Scelidosaurus found in 2000 shows it in its death pose, with all its body armour, including its scutes, in place.*

RIGHT: *The complete Scelidosaurus fossil skeleton has been meticulously prepared and pieced together from where it was found encased in separate rock nodules.*

The question hanging over all these finds is: Why are land-dwelling dinosaurs turning up in marine rocks? All the creatures were found close together and in the same sedimentary layer, so it looks as though this was a family or herd with parents and young that died together in a single event. Scientists are unsure, but perhaps the scelidosaurs were washed out to sea by a tsunami or flash flood, or had been trying to cross a river and got caught up in the current.

For land-based creatures, rivers and seas would have been treacherous places, with the danger of their huge bodies and stout frames being quickly submerged and carried away by the strong currents.

The bodies would have invariably drifted out to sea before sinking to the muddy bottom. The swept-back angle of the legs of the best-preserved *Scelidosaurus* is certainly consistent with its sinking through water prior to a burial on the sea bed. This mystery may never be fully solved, but one day another *Scelidosaurus* may emerge from the Charmouth cliffs, and shed more light on the matter.

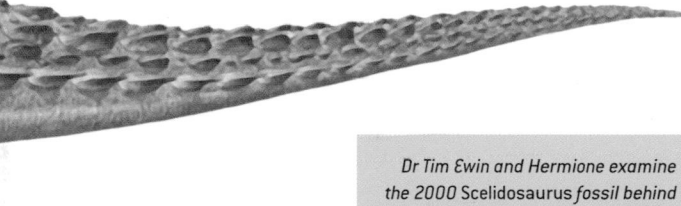

Dr Tim Ewin and Hermione examine the 2000 Scelidosaurus *fossil behind the scenes at Bristol City Museum and Art Gallery, as Tim makes final preparations to put it on display to the public.*

The Wytch Farm Oil Field

FOSSIL FUELS: OIL IN THE SOUTHWEST

The energy we use to power our daily lives – whether it's for lighting, heating, cooking or transport – is almost entirely derived from fossil fuels. These sources of energy were formed from the remains of ancient life that inhabited Earth many millions of years ago. Coal is the compressed remains of vast amounts of plant material that accumulated in swampy environments on land, whereas most oil and gas formed from the decay of countless billions of tiny planktonic algae that lived in the sea millions of years ago. When fossil fuels are extracted from the ground and burnt, they release energy from the sun that was originally captured by the plants and algae during photosynthesis. How long we can continue to rely on fossil fuels is unknown: resources are finite, accessible reserves are in decline and there are fewer places to look for commercially viable reserves that will satisfy our growing demands for energy. In recent decades, British rocks both offshore and onshore have been significant sources of this globally important fossil commodity.

The process of oil formation takes time: in most cases oil forms in rocks such as shales and mudstones that are rich in the organic remains of microscopic marine algae. When these rocks are buried, heated up and pressurized deep underground, a carbon-rich chemical residue called kerogen is all that is left of the original algae. If the rocks are heated up further by being more deeply buried and compressed under layers of younger rocks, the kerogen turns into crude oil, a dark, often treacly liquid. Natural gas can also be found near to these deposits, depending on the exact conditions.

Once formed, the crude oil usually migrates out of the source rock, slowly squeezing its way through the pore spaces of different rock units until it hits an impervious layer. When significant amounts are trapped, the crude oil forms a reservoir – not an underground pool or lake

An aerial view of part of Poole Harbour in Dorset shows Sandbanks and Brownsea Island. Beneath the area lies the Wytch Farm oil field, the biggest onshore oil reserve in Britain, which supplies 25,000 barrels of oil a day.

as the name might suggest, but a rock layer in which the pore spaces are filled with oil or gas, often under enormous pressure.

If oil production in the UK is mentioned, the North Sea immediately springs to mind. Here, huge reservoirs of oil and gas were discovered offshore in the late 1960s and 1970s. The North Sea has meant Britain has been largely self-sufficient in these fuels for decades. But what many people might not realize is that there are onshore oil fields in Britain that have been commercially exploited since the early 1900s.

Oil-bearing rocks underlie parts of the Midland Valley of Scotland, northeast and central England, and many places along the south coast. Today, the giant Wytch Farm oil field in Dorset, discovered in 1973 and operated by BP, is by far the biggest onshore reserve and dwarfs many of the North Sea offshore fields. It is also the largest-known onshore oil field in the whole of Western Europe.

Oil in the southwest of England, like most British oil, originally formed in organic-rich, Jurassic-aged shales, but it migrated, exploiting pore spaces, faults and rock

A beam pump works on Furzey Island in Poole Harbour. From the shore, it is hard to see any evidence of oil extraction, as everything is painted brown and kept below the tree line.

fractures to form underground reservoirs in rocks of various ages. At Wytch Farm, it is extracted from three separate reservoirs that extend under Poole Bay, Poole Harbour and the eastern end of the Purbeck Peninsula.

The most important reservoir is in Triassic-aged Sherwood sandstone. These rocks are found about a mile underground. Estimates of exactly how much oil is trapped in the Wytch Farm reservoirs and how long it will go on producing vary, but there is thought to be about a fifth of the original 500 million barrels of recoverable oil left. Wytch Farm currently produces almost 25,000 barrels of oil a day – a small but significant fraction of our daily consumption as a nation (roughly 2 million barrels a day).

If you have never heard of the Wytch Farm oil field, that could be because it is very difficult to see or hear any evidence of the wells, pumps and other infrastructure required to extract and transport the oil. The reason for this is that the entire oil field is in an Area of Outstanding Natural Beauty, including many Sites of Special Scientific Interest (SSSIs), parts of the World Heritage Jurassic Coast and other protected wetland and wildlife sites. To minimize impact, special 'extended reach' drilling techniques have been used, and acoustic muffling ensures minimal noise. All the drilling rigs and well-site technology is painted brown and kept below the tree line. Everything possible is done to prevent any oil escaping into the environment.

One of the most intriguing places where this style of oil production is taking place is Furzey Island, a small private oasis in Poole Harbour. From the shore it may look like any of the other forested islands in the bay, but since development work by BP in the mid-1980s there have been two well sites for the Wytch Farm oil

A colony of red squirrels lives on Furzey Island. Conservation of red squirrels is increasingly important as they are displaced by grey squirrels throughout Britain.

field cleverly hidden among the trees. Here beam pumps, or 'nodding donkeys' as they are popularly known, and other machinery pump out the oil and pump water back in to fill the void left in the reservoir rocks. Crude oil from Furzey is transported by pipeline to a gathering station on the nearby mainland, before being transferred to a storage facility near Southampton.

The island is a haven for wildlife: owls, peregrine falcons and deer live side by side with the technology. Of particular importance is a colony of red squirrels as Furzey is one of very few places where these animals still survive in southern Britain. Not all of the wildlife is native: brightly coloured golden pheasants were introduced to the island when it was a private home but can still be seen wandering in the woodland. At least here, when the oil eventually runs out, it will be no time at all before the entire island is reclaimed by nature.

DID YOU KNOW? **MICROPALAEONTOLOGY**

Micropalaeontology is the study of microfossils, which are fossilized microscopic organisms generally less than 1 mm in size and often much smaller. Micro-organisms are incredibly diverse, but well-known examples include algae and plankton. They are extremely numerous in the fossil record and most sedimentary rocks will contain several different types: *coccoliths* (see p. 127), *foraminifera*, *radiolarians* and *dinoflagellates* are just four examples of organisms found in marine rocks; others such as diatoms (above) and pollen grains can be found in rocks formed in many different environments.

Despite being so tiny, when magnified, microfossils display an amazing variety of forms, ranging from discs, doughnuts and spiky balls to delicate structures that resemble waterwheels, colanders, nets, footballs and even spaceships. Some species evolved quickly, which means that their evolutionary sequence can be used as a geological clock to date the rock they are found in. It is this potential as a biostratigraphic marker, as well as the environmental information they can provide about a rock's history, that makes microfossils of interest to oil companies.

Drilling for oil is an expensive business, and when an oil company commits to drilling deep underground looking for a new reservoir, it needs to know as quickly and accurately as possible where it is in terms of the underlying rock layers and structures. 'Getting lost' underground is a real possibility without being able to examine the rocks as geologists do on the surface. However, small rock fragments brought up to the surface of boreholes with mud and lubricating fluids will usually contain thousands of microfossils of various sorts. Micropalaeontologists separate them from the rocks using chemical techniques and examine them under a microscope to discern the age of the rock the drill bit is passing through, as well as other information about the rock type and its history. Urgent data can be obtained in a matter of hours, even on an oil rig out at sea. Microfossils may be small, but when it comes to finding oil, they are extremely important.

Carboniferous Swamp

LIFE IN A COAL SWAMP — WRITHLINGTON, SOMERSET

*Exploitation of coal in the Bristol-Somerset region predates the Industrial Revolution.
In this area, the coal occurs in late Carboniferous rock strata, famous for
their fossil plants for nearly 300 years.*

Fossil plants from these coal mines can be found amongst the Woodwardian collection, assembled by the London physician John Woodward (1665–1728), which formed the basis of the Sedgwick Museum at Cambridge University. The last mines were worked right up until the 1970s, and some of the old tips were reworked for coal between 1984 and 1986. The spoil soon attracted fossil detectives as word got around of interesting pickings.

Luckily, the potential of the remaining 3000 tons of spoil led to a concerted conservation effort in which collectors, conservationists and geological and palaeontological groups collaborated for the communal good. The site became a fenced conservation area in 1987 and a protected SSSI, known as Writhlington Geological Nature Reserve, as well as a Regionally Important Geological Site.

Systematic collecting by dedicated amateurs and professional fossil detectives since 1984 has uncovered more than 1200 insects and other animal fossils. This is the largest-known collection of late Carboniferous arthropods in the world, especially of insects from a single locality. Even so, finding fossils here is not as easy as it sounds: even an experienced collector might uncover no more than one insect in a day's work.

The most common insects to be found here are cockroaches (blattodeans): over 700 specimens have been collected to date. Some 70 specimens of extinct mite-like arachnids called phalangiotarbids have also been found, followed by the extinct plant-eating protorthopterans, among others. The total diversity includes several other animals whose remains are only known from a few specimens, such as a single tetrapod footprint.

Fossil detectives swarm over the old coalmine spoil in the hope of finding the fossil remains of the life that originally occupied this 307 million-year-old swamp.

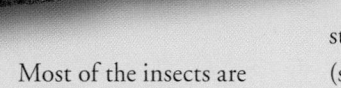

Well-preserved remains of a fossil cockroach were the reward for one lucky prospector's labour on the coal tip.

Most of the insects are associated with the plant debris of swamp-dwelling clubmosses (lycopsids), and less commonly with ferns and seed-ferns (pteridosperms), which grew on the higher and drier river banks and levées, or the horsetails (sphenopsids), which grew between the wet swamps and drier banks.

Some 307 million years ago, Writhlington lay close to the equator and within one of the world's first tropical rainforest ecosystems. Rivers meandered through the forest and seasonally burst their banks onto floodplains and into the surrounding dense vegetation. There were permanent lakes, ponds and swamps with different plants adapted to the varying conditions. Peat swamp deposits built up in the flood basins, but periodic flooding drowned them in blankets of mud and silt. Eventually, plant growth was re-established and formed another peat layer and potential coal seam.

It has been estimated that a metre of mud and silt could be deposited in as little as five years, while it took around 700 years to accumulate 1 metre of peat that would eventually be compressed into 10 cm of coal. Some ten coal seams within 500 metres of strata were worked, and most fossils came from the shales above the oldest, Number 10 seam, although some also come from above the Number 1 seam.

The whole Writhlington ecosystem was founded on the dense vegetation, with the peat swamps being largely occupied by clubmosses, especially *Lepidodendron* and *Sigillaria*. Remains of their stumps have been found, as well as fragments of their leaves (*Cyperites*) and reproductive structures (for example, *Lepidostrobophyllum*). Ferns (such as *Neuropteris* and *Pecopteris*) and seed-ferns preferred the slightly better-drained channel banks and levées. From all these finds it is possible to reconstruct two main fossil environments.

The clubmoss swamp forest and leaf-litter were home to the cockroaches. These incredibly successful insects fed and bred on decaying plant matter, as did rare giant myriapods (*Arthropleura*) that grew to 50 cm long. The cockroaches were accompanied by predatory arachnids such as *Pleophrynus* and *Phalangiotarbus*. The trunks of the clubmosses would have been home to a number of arthropods, but so far only a whip spider (amblypygid – *Protophrynus*) has been linked to the area.

There is also a record of an extinct herbivorous flying dragonfly-like palaeodictyopteran, which had specialized sucking mouth parts for feeding on plants. Some of these flying insects were spectacularly large, with wingspans of up to 55 cm, and are among the largest insects on record. Their wings were strongly patterned, presumably as camouflage to protect them from predatory dragonflies. The top predator is thought to have been a pelycosaur synapsid (a kind of reptile), represented by a single footprint. These animals grew to over a metre in size.

Freshwater animals that lived in the more permanent water bodies include bivalved clams (*Anthraconauta*), clam shrimps (conchostracans), seed shrimps (ostracodes), a horseshoe crab (xiphosuran – *Euproops*) and a possible hybodont shark, whose remains are represented by a fossil eggcase ('mermaid's purse').

Southeast England was the location for one of the most famous fossil frauds in the history of science. In 1912, Charles Dawson announced a remarkable discovery from a gravel pit near Piltdown in Sussex: pieces of an apparently primitive human skull and jaw bone. 'Piltdown Man', as the fossils became known, was identified as an early hominid, officially named *Eoanthropus dawsoni*, and hailed as the missing evolutionary link between apes and humans. Dawson enlisted the help of Arthur Smith Woodward from the British Museum to unearth more fossil evidence from the same site. But in 1953, after years of rumour and speculation, the whole thing was exposed as an elaborate hoax: the original jaw was from an orang-utan, and the skull fragments were no older than Medieval times. Dawson and Woodward were both dead by the time the trickery was revealed. Some people think they were duped by a skilled hoaxer, and the identity of the true fraudster remains a mystery to this day.

Piltdown aside, there are many places of genuine fossil interest throughout the southeast region. The rocks are predominantly Cretaceous in age, dating from about 146 to 65 million

The Seven Sisters on the Sussex coast form part of the distinctive white Chalk cliffs of southern England. The Chalk is the remains of billions of microscopic marine organisms that lived in an ocean that covered most of Britain more than 65 million years ago.

The Chalk

LIFE AND DEATH IN THE CRETACEOUS SEAS

During the late Cretaceous period from about 100 to 65 million years ago, almost all of Britain, apart from a few isolated patches of land in the far north, was submerged beneath a vast warm ocean. Huge numbers of tiny planktonic animals called coccolithophores floated in the water. The remains of countless billions of these microscopic creatures gradually built up a thick carpet of sediment on the sea bed which formed the Chalk, a very pure white limestone and one of Britain's most distinctive rock formations. The iconic white cliffs found along the south coast from Kent to Dorset and on the Isle of Wight were carved from the Chalk and the beaches below them provide rich pickings for fossil detectives.

years ago, and contain abundant evidence of ancient life in this period from both land and sea. The first story in this chapter is about the Chalk; many exposures along the south coast contain fossils ideal for collecting. The region is also the best place in Britain for dinosaur fossils, and we shall look at some recent discoveries, as well as the story of Gideon Mantell, the man who first suggested that giant reptiles once stalked Earth.

Every single lump of Chalk contains an immense number of individual fossil coccoliths, so if you pick even a small piece you will have already amassed a huge fossil collection. But coccoliths are far too small to be seen with a naked eye — they were only discovered after the invention of the scanning electron microscope. What draws flocks of hopeful fossil-hunters to the Chalk is the impressive range of much larger fossils that reveal the diversity of life in the late Cretaceous ocean. Chalk fossils are popular among many collectors because they readily separate from the host rock and are therefore relatively easy to prepare.

Echinoids (sea urchins) are a particular favourite among common Chalk fossils. Echinoids have a hard shell-like skeleton, made of interlocking calcite plates, that fossilizes well. The shape of an echinoid is a good guide to how the animal lived. Rounded sea urchins

The Needles, a line of three Chalk sea stacks, form the westernmost point of the Isle of Wight. A needle-like column between the first and second stack, called Lot's Wife, collapsed in a storm in 1764.

such as *Tylocidaris* lived on the surface of the Cretaceous sea floor, and flattened, oval- or heart-shaped urchins such as *Echinocorys* and *Micraster* burrowed into the soft sea bed. Just like modern sea urchins, the fossil species were covered in spines, but these are rarely found still attached. *Tylocidaris* had heavy, club-shaped spines that were very big in comparison to its body, and the spines are often found as fossils on their own. By analogy with living sea urchins, scientists think the spines were used for defence against predators.

Ammonites of various shapes and sizes are abundant in the Chalk. Coiled species, such as the spiral-shaped *Turrilites*, were common in the late Cretaceous, and other species grew to enormous proportions. Some of the biggest fossil ammonites you are ever likely to see can be found near Peacehaven, in Sussex. The site is protected and it

would be illegal to remove them, but the fossils are so big it would be virtually impossible to anyway.

Embedded in the Chalk shore-platform in front of the sea defences to the west of the town, there are about 200 fossils of *Parapuzosia leptophyla*, a species of giant ammonite that could grow up to 1.5 metres in diameter. Some of the fossils are perched on naturally eroded plinths, which look a bit like ammonite coffee tables and are easy to see at low tide. They have been preserved as casts that formed when sediment consolidated inside an empty ammonite shell that later dissolved away. When alive, these enormous ammonites would have propelled

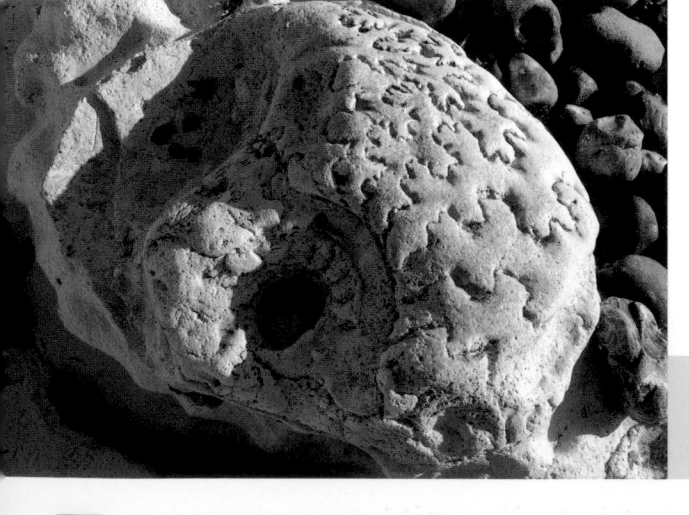

This giant ammonite, Parapuzosia leptophyla, *is over 1 metre across. The frilly lines, or sutures, mark where the internal chambers joined the outer shell.*

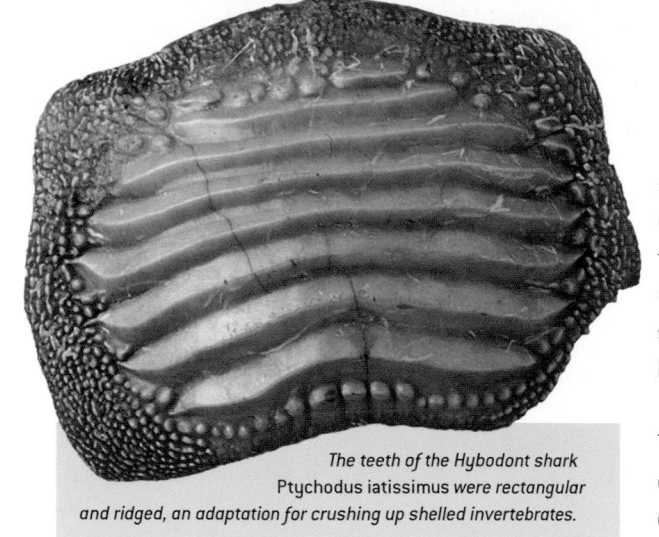

The teeth of the Hybodont shark Ptychodus iatissimus were rectangular and ridged, an adaptation for crushing up shelled invertebrates.

themselves slowly through the clear warm water of the Cretaceous seas rather like a giant turtle swimming in the Pacific today.

Fossil sponges are also common in the Chalk. Sponges are primitive filter-feeding animals that flourished on the chalky sea bed. Some living sponges can survive being pushed through a sieve. Afterwards, the cells regroup and form into a new sponge. Sponges that have a hard skeleton made of fused needle-like spicules fossilize well and in the Chalk they come in an array of shapes and sizes. Most are a few centimetres long and roughly conical or vase-shaped, but others, such as *Laosciadia,* are mushroom-shaped.

As well as many other invertebrates, such as bivalves and gastropods, the Cretaceous Chalk seas were full of larger animals. Keen-eyed fossil-hunters can occasionally spot fish scales that stand out as shiny brown or pink flakes against the white rock. Shark teeth are common because sharks constantly replace rows of old teeth with newer ones, and an individual animal might shed as many as 1000 teeth during its lifetime, which is a lot of potential fossils.

Ptychodus was a type of shark, now extinct, that only lived in the late Cretaceous. It had unusual rectangular-shaped teeth with elevated serrated ridges. The teeth were adapted for crushing up shells of invertebrates

such as bivalves and ammonites, and they make very distinctive and attractive Chalk fossils. Other shark teeth, such as those of *Cretolamna woodwardi*, are a more typical pointed-dagger shape. Fossils of reptiles such as turtles, ichthyosaurs and plesiosaurs are rare but not impossible to find.

The end of the Cretaceous was marked by a mass extinction that completely wiped out about 65 per cent of all living species, including many of those found in the Chalk. Of more than five major extinction events to have struck life on Earth, this one is probably the best known. It was the end for complete groups of animals including dinosaurs, pterosaurs, plesiosaurs, belemnites and ammonites, to name but a few.

The cause of the extinction has been hotly debated but it is likely to have resulted from a combination of factors, including falling sea levels that led to catastrophic habitat destruction; a giant meteorite impact that threw massive amounts of dust into the atmosphere, blocking sunlight for more than six months; and prolonged volcanic eruptions that severely disrupted global climates for many years. It was definitely a disastrous end for many animals familiar to us as fossils, but it was just the beginning for others such as birds, insects and mammals, which took over as the dominant groups in the next geological era.

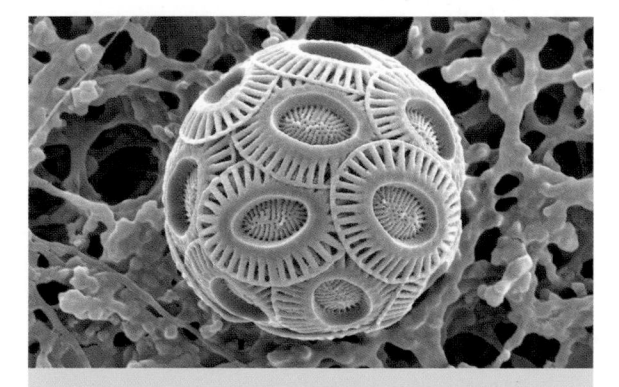

A scanning electron micrograph of a complete coccosphere. Normally the oval plates (coccoliths) separate after death.

Note: Chalk with a capital 'C' refers specifically to the late Cretaceous formation of the very pure, fine-grained, white limestone made from the remains of coccolithophores that covers large parts of southern and eastern England; chalk with a small 'c' refers to a similar rock type of any age.

Gideon Mantell

... AND THE DISCOVERY OF THE DINOSAURS

The Sussex village of Cuckfield played a pivotal role in the discovery of dinosaurs. But the contribution of one man in particular – Gideon Mantell – has gone down in history. Mantell was the first man to identify a dinosaur fossil as belonging to a giant reptile that once roamed the land. His discovery of an *Iguanodon* tooth in Cuckfield, as well as the remains of other dinosaurs such as *Hylaeosaurus*, made him and the village famous. Mantell's discoveries were among the first dinosaurs officially known to science.

Gideon Mantell was born in 1790 in Lewes, about 25 km (15 miles) from Cuckfield. As a child he was an avid fossil collector and accumulated a first-rate collection from the Sussex Chalk Downs. In 1811, he qualified as a doctor and set up a medical practice in his home town. A few years later he married Mary Woodhouse, the daughter of one of his patients. Mary shared his interest in fossils – this was fortunate, because Mantell spent every spare moment he had pursuing his fascination with geology.

Portrait of the Lewes doctor Gideon Mantell, who first realized that bones and teeth from a Cuckfield quarry must have belonged to a giant herbivorous reptile.

A drawing from one of Mantell's books showing the quarry at Whiteman's Green. The spire of Cuckfield church is visible in the background. Nothing remains of the quarry, the site of which is now covered by a football pitch.

The demands of Mantell's job meant that he couldn't go out collecting fossils as often as he would like, so he relied on various contacts to send him interesting finds. In 1820, he received a box of fossils from a quarry at Whiteman's Green in Cuckfield, inside which he found some puzzling fragments of very large bones. Intrigued, he began to make regular trips to Cuckfield, often accompanied by Mary and his children, and it was on one of those visits in 1821 or 1822 that he (or some say his wife) made a momentous discovery: a fossilized tooth unlike any he had ever seen before.

The dark-brown fossilized tooth was about 3 cm long with a very worn grinding surface. Mantell had been studying the mysterious fossil bones from Cuckfield for some time, and this tooth was the final piece of evidence he needed for a ground-breaking theory. Mantell thought that the fossils from Whiteman's Green belonged to an extinct herbivorous reptile, possibly up to 10 metres long, that had lived on land a very long time ago. In effect, Mantell had discovered dinosaurs, although it was some time before they were formally described as such.

Mantell was convinced that his interpretation was right, but he wanted to be sure. He began asking for the opinions of members of London's Geological Society and other

well-respected experts. During this time, he published his first geology book, *Fossils of the South Downs*, with illustrations by his wife. Unfortunately, poor sales of the book left Mantell with a considerable debt and the responses from fossil experts about his giant reptile theory were wrongly discouraging. In a famous error of judgement, the eminent French anatomist George Cuvier dismissed Mantell's fossil tooth as nothing more unusual than the incisor of a rhinoceros.

Although disheartened, Mantell continued to collect more fossils from Cuckfield, refine his ideas and try to get them accepted. One of the biggest hurdles he faced was prejudice from the academic elite. It was all too easy for them to ignore a relatively lowly country doctor, but in 1824, Mantell's luck began to change. Georges Cuvier revised his initial opinion of Mantell's fossilized teeth and agreed they were reptilian. Then, on a visit to the Hunterian Museum in London, Mantell

Some of the original Iguanodon teeth found by Mantell and his wife, which can be seen on display at the Natural History Museum in London. The tooth on the right is 5.3 cm long.

noticed the striking similarity between his fossil teeth and those of a modern-day iguana. Taking advice from the geologist William Conybeare, Mantell named his ancient reptile *Iguanodon* (meaning 'iguana tooth') and finally had a paper accepted by the Royal Society in 1825. Later that year, Mantell was elected as a member of the prestigious society and at last began to receive recognition for his efforts.

Although many people think that *Iguanodon* was the first dinosaur to be discovered, that title actually goes to *Megalosaurus*. While Mantell was working long hours on his Cuckfield fossils, the Oxford University geologist Dr William Buckland, had been studying some large fossil bones of his own from Stonesfield in Oxfordshire. Buckland knew about Mantell's evidence for giant reptiles in Cuckfield and it seems he was anxious to prevent Mantell publishing his theory first. At one point the publications committee of the Geological Society warned Buckland against including some of Mantell's fossil evidence from Cuckfield in one of his own papers. In 1824, Buckland published a somewhat hastily

prepared account of a giant reptile from Stonesfield, a large carnivore which he named *Megalosaurus*, meaning 'giant lizard'. He pipped Mantell to the post, and *Megalosaurus* became the first dinosaur officially known to science. *Iguanodon*, formally described just one year later, is distinguished as the first-known herbivorous dinosaur.

Mantell will always be remembered for *Iguanodon*, but his story does not end there. He went on to identify many more dinosaurs, including *Hylaeosaurus,* which was also based on fossils from Cuckfield. Mantell published numerous scientific articles and books including *The Age of Reptiles* in 1831 and the popular *Wonders of Geology* in 1838. He moved to Brighton, where he opened his outstanding fossil collection to the public as a museum and gave popular lectures on geology. In 1841, Richard Owen, a young and highly ambitious anatomist, placed the giant reptiles *Megalosaurus*, *Iguanodon* and *Hylaeosaurus* into a new group of animals he called the dinosaurs, and Mantell continued to be part of this exciting new field of science.

But for all his later successes, Mantell's achievements were overshadowed by continual frustration with the scientific establishment and academic rivalry, first with Buckland and later with Owen. Mantell's determination to succeed as a geologist to the exclusion of everything else in his life eventually led to the break-up of his marriage and the demise of his medical career. To make ends meet he sold his fossil collection in 1838 for £4000 to the British Museum (less than he had hoped for) and moved to London. After a serious carriage accident Mantell's health deteriorated, and he died in 1852 a rather lonely man.

For almost 150 years after Mantell's death, there was nothing at Whiteman's Green to mark the location as the site of his most famous discoveries, but in 2000, the villagers of Cuckfield clubbed together and erected a monument to Mantell and his ground-breaking ideas.

A modern reconstruction of Iguanodon. *We now know it had two large thumb spikes, as shown here, but when Mantell first found a fossilized spike he thought it was a horn, an idea that stuck for many years.*

The Isle of Wight

THE BEST PLACE IN EUROPE TO FIND DINOSAUR FOSSILS

The Isle of Wight is known as 'Dinosaur Island', and is the best place in Europe to find dinosaur fossils. Dozens of almost complete dinosaur skeletons have been found and up to 20 different species of dinosaur have been identified, some of them unique to the island. There are two main reasons why the Isle of Wight is abundant with dinosaur fossils. Firstly, there are plenty of soft rocks of the right type and right age exposed along the coast. In addition, there are frequent cliff falls and high rates of coastal erosion, which means that fresh material is constantly being exposed.

Isle of Wight dinosaurs all date from the early Cretaceous period. At that time, about 146 to 120 million years ago, the area of land that forms the island today was part of a vast floodplain that stretched from what is now the south of England across the Channel to northern Europe. It was crisscrossed by river channels and was home to many types of dinosaur. The rivers deposited layer upon layer of sand and mud across the open landscape, regularly burying the remains of dead dinosaurs. Those sediments now form a suite of rocks known as the Wealdon Beds, part of which is particularly well exposed at two coastal sections: one in the southeast at Yaverland near Sandown, the other in the southwest from Atherfield to Compton Bay.

The Wealdon Beds are highly significant because few rocks so rich in dinosaur fossils from the early Cretaceous are exposed anywhere else in the world.

Fossils found here give palaeontologists a useful insight into dinosaurs from this period and are an important link between fossils found at older and younger sites in other countries.

By far the most common type of dinosaur found on the island is the plant-eater *Iguanodon*. In fact it is so common that any isolated fragment of bone that is hard to identify is usually considered part of an *Iguanodon*. These creatures travelled in huge herds across the floodplain, and many near-complete skeletons have been found on the island. Two different species have been discovered here so far: *Iguanodon bernissartensis* and a smaller and faster-moving species called *Iguanodon atherfieldensis*, named after Atherfield Point on the southwest coast, where the type-specimen was found. This second species has recently been thought of as sufficiently different from other *Iguanodons* to be classified in a new genus of its own; it is now called *Mantellisaurus atherfieldensis* after Gideon Mantell, who named the first *Iguanodon* in the early 1800s.

Another common find is the dinosaur *Hypsilophodon*. When it was first discovered on the island in 1832 it was thought to be a baby *Iguanodon*, but after further specimens were found scientists realized it was an entirely different species. It grew up to 2.5 metres long and walked around on its larger back legs with a raised tail. It probably used its smaller front limbs for grabbing vegetation. In many ways *Hypsilophodon* resembles a small *Iguanodon*. It also travelled in large herds across the floodplain, but given its much smaller size and more delicate structure, it would have been able to run much faster. Up to 100 individuals have already been found in a single layer of the Wealdon Beds now known as the 'Hypsilophodon bed', and scientists think many more skeletons may yet be found. Despite this creature's ability to run fast, this discovery suggests that a large herd of *Hypsilophodon* was caught out by an

This Iguanodon skeleton, on display at the Dinosaur Isle Museum, was found on the island in 1975 by Steve Hutt. The fossilized bones include the arms, legs, vertebrae and fragments of pelvis. The rest of the skeleton had been washed away by the tide.

enormous flash flood and the creatures were drowned and buried together.

Other types of dinosaur from the Isle of Wight include *Polacanthus foxii*, an armoured dinosaur that grew up to 5 metres long and was covered in spines, and large predatory dinosaurs such as *Neovenator salerii* and *Eotyrannus lengi*. *Eotyrannus* was discovered in 2001 and has not yet been found anywhere else in the world. It was about 6 metres long and probably hunted *Hypsilophodon* and small *Iguanodon*. The range of species and quantity of bones found on the Isle of Wight, particularly on the southwest coast, are remarkable and equal to more famous fossil-finding sites in North America and Mongolia. The diversity of species suggests that some, such as *Iguanodon* and *Hypsilophodon*, lived in the floodplain area. Other dinosaurs that are found much more rarely were probably just visiting the area temporarily.

Since the first dinosaur bones were found on the island in 1832, the Isle of Wight has attracted all manner of

fossil detectives. One of the many resident dinosaur hunters is Martin Simpson. Like other enthusiasts based on the island, Simpson does what he can to salvage the precious dinosaur heritage from the destructive forces of the sea. In his case, this means braving the treacherous cliffs of the southwest coast in all weathers and checking to see if any good fossils have been exposed. In 2004, while searching the Wealdon Beds in Brighstone Bay, the distinctive pink sheen of a dinosaur bone against the background of a red clay layer caught Simpson's eye. He had found what he thinks is probably half a *Mantellisaurus* stuck head – first into the cliff. Simpson has already excavated several vertebrae and part of the hip (which are on display at the island's Dinosaur Farm Museum) and spends as much time as he can at the dig site in the hope of finding more.

As well as dinosaur bones, you can also find dinosaur tracks on the Isle of Wight. At Hanover Point, an area of the southwest coast managed by the National Trust, there are some excellent fossilized casts of *Iguanodon* footprints. Some of the casts have eroded out of the cliff and foreshore and are left as massive three-toed rocky lumps on the beach. They are so big that there is no doubt they were made by the largest known *Iguanodon* species – *Iguanodon bernissartensis*, which was up to 3.5 metres high at its hip. More than 30 virtually complete skeletons of this dinosaur were found in a coal mine at Bernissart in Belgium in 1878. Many arguments about what *Iguanodon* had looked like were resolved after this amazing discovery.

At low tide at Hanover Point, there is also a chance to see some fossilized tree trunks known as the 'pine raft'. Scattered across the foreshore are fossilized sections of gymnosperm logs, up to 1 metre in diameter and up to 3 metres long. Like most of the dinosaur bones and tracks on the island, they come from the Wealdon Beds and help to reveal what the vegetation of the Isle of Wight was like during the early Cretaceous period.

Simpson excavates his fossils by digging and picking away at the soft rock using a spade, hammer and chisel. This is how he found bone fragments of *Mantellisaurus* that had been lying encased in Wealdon sediments for more than 120 million years. Dinosaur digging is a compelling task, but it takes enormous dedication and patience to excavate a complete skeleton. Simpson is determined to uncover the rest of the *Mantellisaurus*, but he thinks it could take him ten years to unearth and prepare the rest of the fossil – and that is only if the sea does not claim it first.

The fossilized cast of an Iguanodon footprint at Hanover Point on the Isle of Wight. They are best seen at low tide after stormy weather, when wave action has stripped away a covering of sand and seaweed.

Baryonyx – 'Claws'
A FISH-EATING DINOSAUR IN SURREY

*In January 1983, William Walker, an amateur fossil collector, was braving the
winter weather in a Surrey clay pit near Dorking when he spotted something
he had never seen before – an enormous, curved, bony claw.*

The claw was nearly 30 cm long and sharply pointed. Walker took it to the Natural History Museum in London to get an expert opinion and was rather surprised when dinosaur experts quizzed him excitedly about the circumstances of his find.

Smokejacks Pit at Ockley, near Dorking, is a privately owned and working brick pit excavated in early Cretaceous Weald Clay deposits of Barremian age (about 128 million years old). The team from the Natural History Museum knew that such a large claw-bone had to belong to a large predatory and carnivorous theropod dinosaur. They were also well aware that, at the time, there were no known complete theropod skeletons of

early Cretaceous age anywhere in the world (or of any age known in Britain). As it was found in a clay deposit, there was a chance that the claw might well be accompanied by other parts of the skeleton.

When the experts visited the pit with Walker, it did not take them long to realize the scope of the discovery. They could see some loose bones in the heavy, sticky clay but others looked to be encased in hard blocks of iron-impregnated siltstone, which had protected them.

Since the site was a working clay pit, the team knew they had to act at once and enlisted the help of fossil experts to excavate the site. It took the team three weeks of hard graft to uncover the skeletal remains. Each bone and block of rock had to be separated from its neighbour, wrapped in a

*Amateur collector William Walker
holding the hooked 'claw' of the new
dinosaur he found near Dorking, Surrey.*

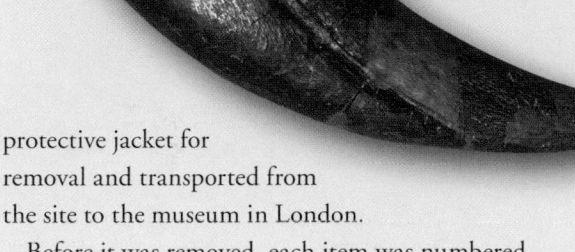

The fossil claw is in fact the bony core on which the even bigger claw, which was not preserved, would originally have fitted.

mounted needles. It was a very time-consuming business that took the best part of ten years.

Reconstructing the skeleton for display was equally difficult. Missing bones had to be modelled, moulded, cast in resin and painted. The scientists had to work out the animal's natural stance – whether it was normally bipedal, quadrupedal or could rise on its hind legs, for example. A framework was then constructed to hold the skeleton in place without damaging it and without looking too obtrusive. Many modern dinosaur displays are made entirely

A reconstruction of Baryonyx walkeri, nicknamed 'Claws'.

protective jacket for removal and transported from the site to the museum in London.

Before it was removed, each item was numbered, measured and mapped to show its original position. There were over 50 blocks, the biggest of which were well over a metre long and very heavy. Many of the bones were of course encased in rock, so the team only had a rough idea of how much of the skeleton remained. Nevertheless, they were pleased because it was such a big animal, stretching over 10 metres on the ground, and it looked as if they had more than 50 per cent of the skeleton, including the skull.

Preparing the skeleton was an arduous task because of the hardness of the iron-impregnated rock. The team used a special industrial air-powered sand blaster and diamond-bladed saws to cut away the excess rock. This work had to be done very carefully because the bone was much softer than the rock encasing it. As soon as any bone was exposed it was coated with latex rubber to protect it from severe damage from the sand blaster. More detailed and finer work was carried out using binocular microscopes, power drills, small chisels and

of moulded bones, which can be drilled and supported internally in dramatic poses.

The dinosaur, later named *Baryonyx* ('large claw'), had a lower jaw shaped remarkably like that of a modern fish-eating crocodile. It had 64 sharp and finely serrated teeth (twice as many as in a typical theropod dinosaur) and was therefore considered to be a fish-eating dinosaur. This prediction was confirmed when the scales and teeth of the late Jurassic-, early Cretaceous-age semionotid fish, *Lepidotes*, were found in the body cavity. The claw that Walker had found was actually the bony core over which an even bigger claw grew. It is presumed that it was a thumb claw that was used to hook fish from rivers, as grizzly bears do today.

The East of ENGLAND

In the east of England, the rocks and fossils become progressively younger the further east you travel. Jurassic limestones and clays underlying the Peterborough area give way to a wide belt of Cretaceous Chalk in East Anglia. Finally, overlying the Chalk and along the eastern coastline of Norfolk and Suffolk there are some of the youngest rocks and sediments in Britain. They date from the Quaternary, the most recent period of geological time, and record the dramatic environmental changes of the past 2 million years in great detail. Within them is a rich fossil record of how life adapted as the climate switched between cold and warm phases of the Ice Age.

During the past 500,000 years, large rivers fed by melting ice sheets that covered the more northerly parts of Britain deposited thick layers of sand and gravel over much of East Anglia. Within the gravels, rather surprisingly perhaps, are a wide range of fossils. They mainly come from the underlying Chalk and Jurassic rocks and were initially eroded out of their bedrock by ice before being redistributed by rivers. It is common to find fragments of belemnites, oysters, sponges, ammonites and

The spectacular cliffs at Hunstanton on the Norfolk Coast. White chalk sits on top of a layer of red chalk that overlies the Carstone, the lower, brown rock layer. The boulders on the beach are a good place to hunt for fossils embedded in the red and white chalk.

echinoids as well as a few more recent fossils including mammoth teeth and tusks. The beauty of fossil-hunting in gravels is that you do not need to travel to East Anglia – the gravels are quarried and end up on paths and driveways across the country. The first story in this chapter is also about fossils redistributed thanks to ice sheets: the amber time capsules that was washed up on the beaches of East Anglia.

Amber

THE FOSSIL TIME CAPSULES WASHED UP ON EAST ANGLIA'S BEACHES

You should keep your eyes peeled when walking along the east of England coast – if you are lucky you might come across pieces of amber washed up on the beach. Many people associate amber with jewellery but it is not a gemstone, it is actually fossilized tree resin. Amber feels warm to the touch, is lighter than expected and can be any colour from dark brown to bright orange, red, yellow or even white. Beach amber is usually scuffed and has a pitted surface after tumbling about in the sea, but once polished, it gleams beautifully.

The amber that washes up on beaches along the east coast did not originate in British rocks, but comes from an area around the Samland Peninsula on the Baltic Coast. Here, amber is found concentrated in a particular layer of sediment known as the Blue Earth that crops out just below sea level. To get at the amber, people have dredged the sea bed and mined the Baltic coastline for centuries. Commercial open-cast mining still operates today, but in the past, storms, floods and glaciers have picked up pieces of amber and spread them across the entire region as far as the North Sea. The amber found on East Anglia's beaches has probably been loosened from the sea bed during storms. Fishing trawlers have been known to dredge up big, almost football-sized chunks of it encrusted with seaweed.

Baltic amber is mostly about 40 million years old, dating from the Eocene period. At this time, the area to the south of Samland was covered in a dense, subtropical forest. A special set of conditions is needed to form amber and only certain trees secrete suitable resin, but in this forest they did so in copious amounts.

The resin naturally solidified and a chemical process called polymerization began. This is a crucial part of the amber-forming process, when the resin becomes like a hard plastic. Lumps of this hard resin were then washed out of the forest by floods and buried in sediments where, over time, they became fully fossilized amber.

Amber has been highly sought after throughout history, and precious beads and carved objects have been found at archaeological sites up to 10,000 years old. A remarkable amber cup was found in a Bronze Age grave in Hove, Sussex, and is probably evidence of trade links between Britain and the Baltic more than 3500 years ago. In eighteenth-century Prussia, an entire room was elaborately decorated in Baltic amber and gold panels. The gold and amber were stolen by the Nazis during the Second World War and then lost and probably destroyed. The room has been painstakingly recreated from photos in the Catherine Palace near St Petersburg.

The pier at Southwold on the Suffolk coast. After stormy weather, pieces of amber can sometimes be found here washed up on the beach. The Amber Museum and Shop in town is a good place to find out more about this intriguing fossil.

Amber is not just prized for its beauty or historical importance, it is also highly sought after by palaeontologists because of what it contains. Amber is known as the fossil 'time capsule' because inside each golden nugget are little bits of ancient life that got caught up in the sticky resin when it first oozed out of a tree. An insect crawling along the bark of the tree might have been entombed and fossilized along with the lump of resin. Not every piece of amber will contain an inclusion visible to the naked eye, but there is a huge range of possibilities. Everything from bacteria, mould, pollen, moss and lichens to leaves, twigs, worms, hairs, flies, ants, spiders and even perfectly preserved frogs and lizards have been found in amber of different ages and from various parts of the world.

Small insects rarely get preserved as conventional fossils, but thousands of extinct species have been identified in amber. Any small creature found in amber was usually still alive when it got caught in the resin, so its behaviour can sometimes be studied. Parasitic relationships, eating habits and even ants trying to rescue one another from a sticky end have all been inferred. One of the oldest insect-bearing ambers ever found comes from the Isle of Wight. Such ambers are much rarer than Baltic amber and was not washed up on the beach. It was found *in situ* in some freshwater sedimentary rocks. Inside was a perfectly preserved spider, almost 130 million years old.

Old amber that dates from more than 65 million years ago, when the dinosaurs were alive, begs the question as to whether *Jurassic Park* could ever become a reality. In the famous film based on the novel by Michael Crichton, scientists

recreated living dinosaurs from their DNA recovered from blood-sucking mosquitoes preserved in Jurassic amber. Sadly, such a feat would not be possible in real life. There are lots of reasons for this. Not only are blood-sucking insects in amber of any age extremely rare (with none being known from the Jurassic), DNA almost certainly cannot survive that long. But this does not make finding a piece of amber while walking on the beach any less exciting. If you are lucky enough to find some amber on the Suffolk or Norfolk coast, you will be holding a fragment of a 40-million-year-old forest in your hand and perhaps even some of the forest's tiny inhabitants frozen in time.

These pieces of beach-washed Baltic amber were collected from the East Anglia coast. The fossilized resin originated in subtropical forests that grew in the Baltic region about 40 million years ago. The largest piece is 2.5 cm long.

A piece of amber originally found in East Prussia, polished to make a pendant, reveals a fossilized fly, Chrysopilus, perfectly preserved inside. This amber came from trees that lived 30 million years ago.

Leedsichthys

A GIANT JURASSIC FISH FROM PETERBOROUGH

In today's oceans, the biggest animals survive by preying on the smallest. The secret to getting really massive if you live in the sea is to spend your life filter-feeding tiny sea creatures such as plankton and krill. The blue whale, the largest animal that has ever lived, and giant fish such as the whale shark and the basking shark, all filter-feed in this way. The fossilized bones of an enormous fish called *Leedsichthys*, which lived around 155 million years ago, tell us that the same feeding patterns existed in the Jurassic seas. Some think *Leedsichthys* could be the biggest fish that has ever lived. Most of the evidence we have to test this claim has been found in the east of England.

Alfred Leeds, a gentleman farmer from Eyebury near Peterborough, collected fossilized bones of Jurassic marine reptiles from nearby brick pits throughout the last part of the nineteenth century. In 1886, he showed some of his new finds to visitors from the British Museum (Natural History) in London, which later became the Natural History Museum. His discoveries included large, platy bones that were originally thought to be the armour of a stegosaurian dinosaur. It was only when Sir Arthur Smith Woodward, an expert in fossil fish, saw the bones that they were identified as the head-bones of a giant fish. Woodward was also intrigued by some of the accompanying bones, known as 'gill rakers'. Filter-feeding fish have numerous long, thin gill rakers in their mouths alongside a mesh for sifting plankton from seawater. The fossil gill rakers that Alfred Leeds had found were like those of a filter feeder but were several centimetres long, whereas in most fish they are much smaller. These fossil bones must have belonged to something very big and very out of the ordinary. Woodward named the new discovery *Leedsichthys problematicus*, because it was going to be hard to work this one out.

The best fossil specimen of a *Leedsichthys* that has been found anywhere in the world was dug out of another disused brick pit near Peterborough over many months in 2002 and 2004. Over 2300 bones, all

An artist's impression of Leedsichthys swimming in the Jurassic ocean 155 million years ago. Fossil bones of this giant fish have been found in brick pits in the east of England since the 1880s.

belonging to one creature, were excavated. Dr Jeff Liston from the Hunterian Museum at the University of Glasgow was the dig leader for this epic excavation.

As Jeff explains, the geology of the Peterborough area is dominated by near-surface exposures of Oxford Clay. The dark grey clay has a high carbon content, which is ideal for brick making, and has been the basis of a large-scale brick industry in the region for decades. In these modern brick pits, huge machines in immense open quarries scrape away at exposed faces more than 30 metres high. The clay was originally deposited year after year at the bottom of an ocean that covered large parts of Britain during the Jurassic period.

The giant *Leedsichthys* was found at Star Pit. This brick pit is now disused, and there's not much to see today.

All that is left is a clay platform gradually being colonized by weeds, but a few years ago this was a hive of activity. In 2001, two students on fieldwork from Portsmouth University spotted some fossil bones protruding from the ground near the base of a 20-metre-high cliff of clay. On closer inspection, bones could be seen sticking out over an 8-metre section of the quarry face – there was definitely something worth digging up in there. By the following year, a team had been assembled, permissions obtained, funds raised and the dig could begin.

After clearing away thousands of tonnes of overburden, Jeff and his team found fossil bones scattered across an area of 25 metres by 10 metres. The bones corresponded to just over half a giant *Leedsichthys*. Many of them were paper-thin and easily fractured. The excavation was an extremely delicate operation, but to the delight of everyone involved, this was undoubtedly the best fossil specimen of the enigmatic fish ever found. The team nicknamed their discovery 'Ariston', after a television advert at the time, because the bones really did go on and on and on and on.

So what was *Leedsichthys,* and just how big did it get? This unique fossil belonged to an extinct group of fish called the pachycormids, meaning 'thick-bodied'. It was the first of the gigantic planktivores, or suspension filter-feeders, that have been found in the fossil record. Up until that time, suspension-feeding fish had not grown any larger than half a metre in length. The elevated organic content of the Oxford Clay indicates that our *Leedsichthys* was swimming around in a highly nutrient-rich ocean. Jeff thinks that there was probably some sort of environmental change in the Jurassic period that led to plankton being more abundant, allowing such a big plankton-feeder to evolve.

No complete specimen of *Leedsichthys* has ever been found, so it's hard to say exactly how big it got. But by comparing body parts with related fish and scaling up to an overall length, scientists have variously suggested between 9 metres and 30 metres. Newer estimates are around 14 metres. This undoubtedly makes *Leedsichthys* the biggest bony fish of all time, but that's not the same as saying the biggest fish of any

The Leedsichthys dig site in Star Pit (disused) near Peterborough. Many people were involved in excavating over 2300 bones from the site between 2002 and 2004.

Hermione and Dr Jeff Liston in front of a freshly exposed face of Oxford Clay in Bradley Fen Quarry. Fossilized sea creatures are sandwiched between the layers of Clay.

kind. Some fish, such as the modern-day basking shark, are in a separate group known as cartilaginous fish, and some scientists believe that cartilaginous species could grow even bigger.

Only about six other significant specimens of Leedsichthys have ever been found. As Jeff explains, the fossil is uncommon because a dead Leedsichthys settling on the sea bed would have attracted scavengers from far and wide. It had big front fins that may have propped it up so it would have taken longer to be buried than if it was lying on its side, leaving even more time for scavengers to devour it and destroy the skeleton. Another reason also relates to its immense size. Any big fish needs to work at weight-saving to help with buoyancy and to save on energy expended while swimming. Many of this fish's bones were not ossified, meaning they were not hard like bones in a human skeleton, but were soft and lightweight. Many of its bones were also porous and very thin. All this means that Leedsichthys fossils are very fragile and break easily when palaeontologists try to collect them. Modern quarrying techniques mean that many specimens have probably been lost before anyone had a chance to spot them.

It is possible that another even bigger Leedsichthys specimen will be found in the brick pits around Peterborough, and one day it may be crowned 'biggest ever fish' unequivocally. But Jeff believes that Leedsichthys does not need to be the biggest to be special. He is now overseeing the preparation of thousands of bones with the help of a team of dedicated volunteers at Peterborough Museum. It will probably take years to fully interpret the skeleton but ultimately it will help us to discover much more about this intriguing giant fish.

A complete fossil tail fin of a Leedsichthys found by Alfred Leeds in 1898. The tail is 2.74 metres from tip to tip.

The West Runton Elephant

A 650,000-YEAR-OLD STEPPE MAMMOTH FROM NORFOLK

Norfolk residents Harold and Margaret Hems know that the best time to look for fossils on the coast is after a storm. In December 1990, when the couple went to check West Runton beach after some bad weather, their fossil-finding tactics were rewarded. Sticking out of the cliff was the pelvic bone of a giant elephant. The Hems alerted Norfolk Museums Service and over the next few years more bones were rescued from the same spot. In 1995, a full-scale excavation took place and, by the end of the three-month dig, a record 85 per cent of the skeleton had been recovered. Known as the West Runton Elephant, it is the world's most complete fossil of a steppe mammoth (*Mammuthus trogontherii*), one of the biggest elephant-like animals that has ever lived.

Nigel Larkin, a palaeontologist at Norfolk Museums and Archaeology Service, had the unenviable task of cleaning and conserving the bones – unenviable because the bones were enormous, very heavy and extremely fragile. During the excavation the bones were carefully wrapped in tissue paper and foil and encased in a removable plaster of Paris jacket. Each thigh bone (femur) alone is a massive 1.5 metres long and, without the support of the plaster cast and extra splints, they could have easily been broken in transit. The skeleton is between 600,000 and 700,000 years old, which means the bones have lost their strength and flexibility compared to fresh bone, but have not been buried long enough for mineralization to take place. As Nigel joked,

if the skeleton had been left in the ground for another 50 million years, it might have been fossilized completely and easier to work with. But then again, it was rescued at an opportune time – having been exposed, the fossil would certainly have been destroyed by the sea within a decade.

When it was alive, the West Runton Elephant was an intimidating beast. From the skeleton we know it was a male that weighed about 10 tonnes and was 4 metres tall at the shoulder, roughly the same height as that classic yardstick, the double-decker bus. This is considerably bigger than the woolly mammoth (*Mammuthus primigenius*) that lived in Britain more recently during the Ice Ages, or any African or Asian elephant alive today. Of all the land-dwelling animals that have existed, only the very largest sauropod dinosaurs would have been bigger or heavier. Like modern elephants, the West Runton Elephant lived on a diet of grass, herbaceous shrubs and other vegetation.

The mammoth's bones were found on the north Norfolk coast, where the cliffs are made from a sequence of sediments that record the changing climates of the

The beach and cliffs at West Runton on the north Norfolk coast where the fossil mammoth was found. The fossils in the cliff sediment record the changing environments of eastern England during the Ice Age.

A reconstruction of the enormous West Runton Elephant, which was a steppe mammoth. It stood 4 metres tall at the shoulder, and this species is possibly the largest land-dwelling mammal ever to have lived.

past million years or so. The layer where the mammoth was found is called the West Runton Freshwater Bed, a dark band that is about 1.5 to 2 metres thick near the base of the cliff. For the past 150 years, this layer has been famous for turning up fossil bones and teeth of big animals such as deer, rhinos, wolves and bears, but it is very rare to find even a few associated bones, let alone a complete skeleton. The layer is made from deposits laid down by a slow-moving river in a climatic environment almost identical to ours today. The layer is particularly rich in a wide range of fossils – within minutes you could find a fish bone, some plant material and lots of small snail shells, for example.

The big excavation in 1995 was a chance to study the smaller fossil content of the bed in detail. Ten tonnes of sediment were collected, sieved and analysed and a huge diversity of life was identified, including various

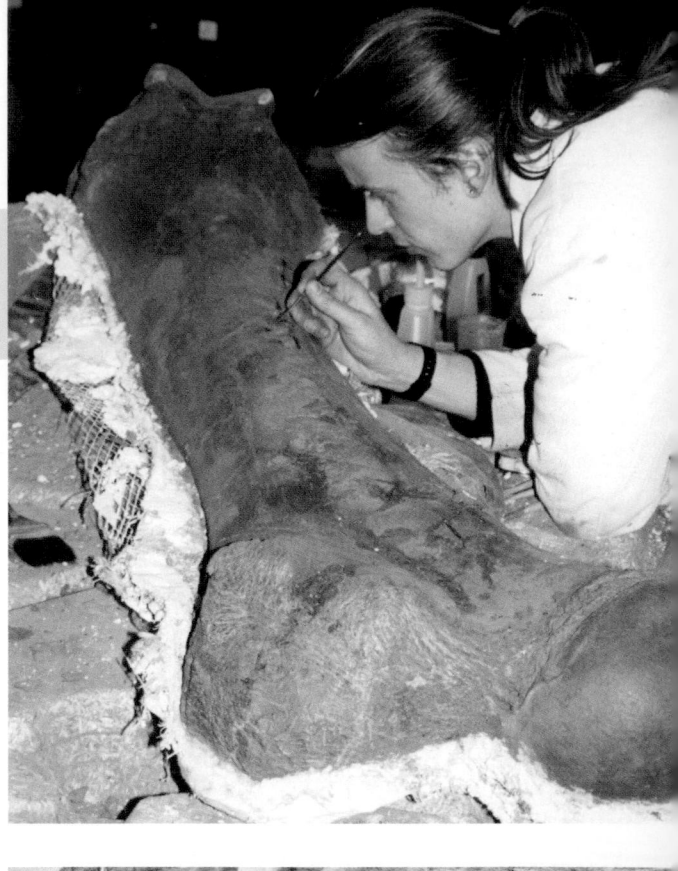

Nigel Larkin painstakingly cleaning one of the enormous bones of the West Runton Elephant. An impressive 85 per cent of the skeleton was found, making it by far the most complete fossil of Mammuthus trogontherii *ever discovered.*

trees and grasses, invertebrates such as snails and beetles, and fish, frogs, newts, lizards, snakes, moles, voles, shrews, beavers, weasels, cats, monkeys – the list goes on. We know from the environmental requirements of the flora and fauna that climatic conditions were not much different around 650,000 years ago compared to now. Although extinct, it seems a steppe mammoth would be quite at home in the Norfolk Broads today.

The West Runton Elephant is such an interesting story because, unusually for a fossilized animal, its death can be pieced together in glorious and gory detail. Its teeth tell us that this creature did not die of old age – it died prematurely at about 40 years old, although it could have lived until about 70. After the mammoth died, one of its tusks was crushed by something big and circular and some of the bones were moved around. Other mammoths had visited the carcass and inadvertently trampled on their dead companion as it lay in the shallow, slow-moving river. Behaviour like this is seen in modern elephants, which are drawn to the remains of their relatives.

The missing 15 per cent of the skeleton consists mostly of the toes, some ribs and the tail, and there are signs of gnawing and scratching on some of the larger bones. Fossilized hyena droppings (coprolites) were found at the site, some even resting on the bones. Hyenas had clearly scavenged on the carcass and left their mark. From the position of the bones the elephant seems to have died lying on its chest, so it may have suffocated or drowned due to the pressing weight of such a huge body. But why didn't it just stand up? Why did it get stuck in the river bed in the first place?

Nigel Larkin and Hermione with two of the rear leg bones showing the damaged knee on the right. Gnawing marks on the bones were made by hyenas. A modern hyena skull is on the table for scale.

We know that an early species of humans called *Homo heidelbergensis* lived locally at a similar time. The ancestor of the Neanderthals but not an ancestor of modern humans, *Homo heidelbergensis* made sophisticated tools and was a fierce hunter, but a steppe mammoth was simply too big a beast to have been hunted, so we can rule humans out.

Careful analysis of the bones revealed a serious knee injury on the back right leg. At some point, the lower shin bone (tibia) had sheared sideways, perhaps in a fall or a fight, but the mammoth had survived for up to two years afterwards. New bone had grown around the joint to compensate for the damage, but this giant beast would have had a bad limp and may well have been in considerable pain.

It would have found it very hard to get about, particularly walking uphill. So perhaps it was the elephant's bad leg that meant it could not get out of the river bed after it had gone down for a drink. It may have got stuck and eventually keeled over and died, and the carcass was slowly buried.

It seems rather a sad tale, but this story contains an important message that has been a running theme of *Fossil Detectives*. The tale of the West Runton Elephant exemplifies the alliance of professionals and amateurs that makes such exciting discoveries possible. It is the combination of keen-eyed children, adult enthusiasts, professional amateurs and trained palaeontologists all working together that has made some of the most significant fossil discoveries in Britain. This is a remarkable feat, and long may the tradition continue.

During the excavation, many of the large bones were wrapped in hard plaster-cast jackets before being moved. Here, part of the skull and one of the tusks has been bolted to a metal frame for extra stability before it is winched up and on to the back of a lorry.

Flint Tools

DISCOVERY OF THE FIRST 'BRITS'

East Anglia provides a wealth of fossil evidence about the animals that lived during the past 2 million years. Very few ancient human remains have been discovered, but stone tools found buried have taught us a great deal about the early human colonization of Britain.

Historically, the most important find of early British colonization was made at Hoxne in Suffolk in the 1790s. Workmen digging out sand and gravel came across a strangely fashioned flint. Perhaps hoping for a reward, they passed it on to the local member of parliament and High Sheriff of Norfolk, John Frere (1740–1807). Frere was a typical country gentleman of the time. Cambridge-educated, he was interested in all manner of antiquities and was a fellow of the Royal Society.

Frere saw that the flint had been carefully fashioned into a sharply pointed and flattened pear-shape with two cutting edges. He realized that the shape was man-made and he also knew it had been dug from several feet below ground, buried within sand and gravels that often contained the fossil bones of extinct Ice Age animals such as the mammoth. Frere prepared a paper that was eventually published in 1800 in the journal of the Society of Antiquaries, in which he described the circumstances of the find.

He suggested that the flint was a hand axe dating from 'a very remote period indeed'.

Frere's paper about the coexistence of humans and extinct Ice Age animals appears to have been the first published, but despite its revolutionary content, his work went largely unnoticed. In the prevailing cultural, social and intellectual climate of the time, speculation about the history of mankind was taboo if it did not fit with tales from the Old Testament. The overwhelming evidence for a human antiquity that stretched back to the Ice Ages was gradually accepted by scientists, but only widely accepted by the public at the end of the nineteenth century.

A direct descendant of Frere had better luck in convincing the public of human antiquity. Mary Leakey's painstaking fieldwork in Africa with her Kenyan husband Louis brought about most of the couple's famous finds, such as *Australopithecus boisei*, *Homo habilis* and the 2.6-million-year-old fossil footprints at Laetoli in Tanzania.

Erosion of the East Anglian coastal cliffs reveals Ice Age fossils and occasionally flints shaped by humans who lived alongside the extinct animals of the Ice Age.

Dated to around 700,000 years ago, these worked flints, found within the fossiliferous cliff sediments at Pakefield, are the oldest evidence for human habitation of the British Isles.

The most exciting discovery relating to the human occupation of the British Isles was recently made by two amateur fossil detectives on the Suffolk coast at Pakefield near Lowestoft. Here, winter storms cut into the low cliffs, eroding their unconsolidated glacial deposits and causing the cliff to retreat inland. Newly exposed glacial sediment is constantly falling onto the beach and being washed away by the tide. Over the years, numerous bones of animals such as the rhino, mammoth and straight-tusked elephant, bison, scimitar-toothed lion, bear and wolf have been found – but there have been no human remains.

However, now and again flint flakes have been found that look to have been purposefully struck from larger flint cores by human hand. These flints were found not in the cliff deposits but already washed onto the beach, making their relative age difficult to determine.

Luckily, some local collectors were prepared to work in all weathers with experts from the Natural History Museum in London to carefully excavate the cliff sediments. They focused on fossiliferous clay layers that lie under the glacial deposits containing large animal bones. The clays contain plant fragments of some 150 species, including water chestnut, broom crowberry, brittle waternymph and floating fern,

along with nearly 100 species of molluscs and beetle-wing cases.

Knowing the temperature tolerances of those extant plants and insect species, it is possible to estimate that summer conditions were warmer than those of today. The presence of hippo remains shows that winter temperatures were above freezing, so the British climate was Mediterranean with hot, dry summers and mild, wet winters. The fossils were evidence for a marshy landscape that was home to many browsing and grazing animals, but the question remained – were any human predators around?

Eventually, 32 worked flints were found at the site, including a worked core and some retouched flints, all made of fresh black flint and in sharp condition. Compared with John Frere's sophisticated hand axe, these were primitive stone tools that predated those made by *Homo sapiens*. Indeed, these 'Pakefield deposits' have been dated at around 700,000 years old – some 200,000 years older than any other British human-related evidence. The crude flint tools could not have been made by Neanderthal people, but probably their evolutionary predecessors, *Homo heidelbergensis*.

Carefully fashioned from flint, this hand axe, like those found by John Frere, was made by extinct human species such as the Neanderthals.

How to collect and look after your fossils

The urge to search for fossils and make a personal collection of these fascinating remains of past life is an ancient one and has been driven by very different needs over the millennia.

Even today, motives for collecting fossils range from simple curiosity to the desire to test a scientific theory and a way of earning a living. Amateur and professional fossil-collectors have always made significant contributions to the scientific investigation of the history and evolution of past life. Inevitably, there are also some potential areas of conflict, especially over access to fossiliferous sites that are of particular scientific interest and value. In Britain, legislation has sought to resolve some of these problems by providing legal protection for selected Sites of Special Scientific Interest. The pursuit of fossil-collecting also raises some important safety issues. A few well-known palaeontologists have even died as a result of avoidable accidents.

For centuries, fossils have been collected and bought purely for qualities such as their appearance, rarity and value. There is still a lively trade in fossils, with significant sales over the internet and from countless stalls in markets throughout the country and abroad. Generally, such fossils come with very little information about where they come from or what age they are. Closer examination often reveals that they have been 'improved' by various techniques, such as painting, additional carving, glueing and cementing, all employed with differing degrees of expertise. The fudging and faking of fossils is an ancient craft no different from that employed on pieces of art or antiques, so be wary before spending significant sums.

A lot of money is paid for prize fossil specimens, especially entire dinosaur skeletons, which may be of

Polished pebbles of cephalopod limestone from North Africa with sections through orthoconic nautiloids are common in rock shops today.

considerable scientific importance. Even these can be subject to unscrupulous practices, however. While many private collectors do make their specimens available to the scientific community, there are others that do not. Even major museums can have difficulty in raising sufficient money to buy really important specimens.

Although it might be difficult for the amateur to find fossils as well preserved as those bought on a market stall, there is a very different kind of satisfaction attached to finding your own. There is also a real possibility of contributing towards the study of fossils, and many professional fossil detectives are glad to cooperate with responsible and dedicated amateurs.

Location, location, location

For fossils to be of any scientific value, they have to come with an appropriate 'pedigree' and 'guarantee of authenticity'. It is essential to know where the fossil has come from, with sufficient accuracy to be able to revisit the exact spot to the inch. Today we have advanced mapping techniques and GPS, but even a GPS fix may not suffice if the fossils lie within a vertical sequence of strata. It is important to know which layer of sediment the fossil came from, so that further collections and information can be gathered. The original orientation of the fossil may also be relevant, so a field sketch, note or photographic record is important.

Finding fossils

One of the easiest ways to find fossils is to let nature do the hard work for you and to search among rocks that have already been broken apart by natural physical processes. Keen fossil-collectors take advantage of winter storms to search beaches as soon as possible for newly exposed and undamaged specimens. This was how famous collectors of the past, such as the Anning family of Lyme Regis in Dorset (see pp. 113–15), managed to find some of their best specimens. It was a hard life, racing to beat the tide, often in bad weather, to recover some heavy slab of rock and drag it to safety.

If you search among already broken rocks, you can also get an idea of the kinds of fossils present in nearby rock strata and, importantly, which fossils are associated with which particular rock types. If a collector is interested in a certain group of fossils, the rock type (lithology) in which it is most commonly found can be targeted for further searching.

One disadvantage of picking up loose fossils is that you can never be certain of their context. One of the few dinosaur fossils from Ireland was picked up on a beach and may well have originated as ballast that had been thrown overboard from a passing ship. The golden rule is that, if the fossil and the rock containing it look different from any of those exposed in the nearest outcrop, be wary.

The preparation of fossils is an exacting and often prolonged affair that requires some specialist tools, infinite patience and considerable expertise.

Careful extraction

One of the most surprising discoveries for the novice collector is the fragility of many fossils. Although fossils are generally made of mineral material, great care needs to be taken when extracting them from the rock matrix. Many specimens have been seriously devalued, in all senses, by careless handling.

Rocks are heavy and it is not uncommon to find a good fossil embedded in a rather large piece of rock. The temptation is to trim away as much of the surrounding rock from the fossil as possible, but there are risks in doing this. Different rock types break in different ways, and it requires experience to predict how a rock might fall apart. Even the most experienced fossil detectives will have inadvertently broken a specimen at some time or other.

Many fossils are only revealed when a sedimentary rock is split open along a bedding plane. In these circumstances, it is common for some parts of the fossil to be left attached to each part of the slab. There is a tendency to throw away the slab that has little on it, but remember that some fossils frequently break apart into two halves. For potentially important specimens there is often significant information that can be recovered by retaining both parts.

Wrap it up

Fossils should be kept in the same state of preservation in which they were found, to avoid important information being lost. That means making sure they are not damaged in transit. Nothing can be more frustrating than carefully extracting a delicate fossil only to find it in pieces when you arrive home.

Pocketing an unwrapped specimen can result in abrasion or scratching, for example, by coins or keys. Continual handling leads to long-term damage from rubbing. To safeguard your fossils, wrap them up carefully. If the fossils are particularly delicate they should be protected with tissue paper and bubble wrap. Tape up your package and include some form of identification. Sometimes, fragile fossils have to be consolidated before they are removed from the rock matrix. These techniques require considerable experience and knowledge of the right kind of chemical solutions for the job in hand.

Fossils such as the carbonized remains of plants, graptolites or insects, and anything with an ornamented surface, can very easily be damaged.

The basic fossil-collector's field kit consists of a hard hat, gloves, safety goggles, geological hammer, hand lens and notebook.

Even careless packing of fossils can result in damage if heavy rocks are jostled together while they are carried. Some specimens may require cotton wool and boxes to protect them.

Write it down

Each specimen must be clearly connected to its original location by some numbering or lettering system that relates to field notes and a locality map. The aim is to provide enough information to be able to relocate each find-spot as exactly as possible and to know how successive fossil finds were related to one another in the field.

It is possible to find more than one element of the same original fossil that was dislocated after death and separated before being buried and fossilized. This is particularly true of skeletal material, especially of large animals whose bones may be scattered by scavengers before burial. If the elements are found at exactly the same horizon they may indeed be related, but if they come from different horizons this is much less likely and one should be sceptical about any possible association. When collecting from a sequence of strata, it is important to know the relative ages of all the fossils retrieved in relation to one another.

Preparation and display

Preparation of specimens is largely a matter of practice, patience, having the right equipment and some manual dexterity. Expert preparation of very delicate fossils is difficult and not a skill everyone can possess. It is worth having a go – but not on your best specimens until you are confident that you will not lose important information or damage the specimen, or yourself. It is better to have an unprepared specimen than one that has been ruined by bad preparation.

Cliffs are potentially dangerous and can collapse without warning, as did this one along Dorset's World Heritage Jurassic Coast.

Be prepared

For any outdoor activity that takes you off the beaten track, it is essential to tell someone where you are going and when you will return. Mobile phones are very useful safety aids and their use has helped to resolve many potentially difficult circumstances. Make sure that yours is charged up. Get an appropriately scaled map, make sure you know how to read it, and carry a compass.

Suitable clothing for all eventualities is a must. The British weather is notoriously fickle, and proper rainproof and warm clothing is required for most of the year. Good, strong boots are also needed – rocks are heavy, feet are easily damaged, and trainers provide no protection. If you are going anywhere near cliffs or rock overhangs, a safety helmet is essential. If you have to do a lot of hammering, a pair of mason's leather gloves are useful and even then a selection of plasters will come in handy. Bruised and battered fingers are an occupational hazard in this business.

Dedicated fossil-hunters might invest in the standard hardware: rock hammer, chisel, hand lens, field notebook, compass, wrapping material and locality map. All this requires a strong backpack. Check the stitching on the shoulder straps, as the straps of cheap backpacks can sometimes tear when any weight is placed in the pack.

Countryside code

Most of the land in the British Isles is privately owned and unfortunately, unlike in Scandinavia, one cannot just go wherever one pleases to collect fossils. Always make sure you have permission to start looking for fossils on private land. In many rural areas this is not a problem and landowners may be no more than curious to know what you find. However, this is not always the case and there are many areas where over-collecting and irresponsible collecting has turned landowners against fossil-hunters. Remember the countryside code and its common-sense rules about gates, fences and livestock.

Many of the most accessible and best-exposed areas of fossiliferous rock are along our coastlines, much of which can be accessed freely but not always safely. Quarries, road cuttings, railway cuttings and rock cliffs are all places where rock strata are well exposed and may look promising for fossil-collecting, but they are all hazardous. Rock falls and slope collapses are unpredictable. It is always surprising how many people fall foul of these dangers each year – do not be one of them.

Glossary of fossils

Ammonoids – class Cephalopoda, phylum Mollusca

❦**Where are they found?** Widespread in marine sediments, best preserved in some shales and limestones.

❦**What age are they?** Early Devonian to end of the Cretaceous.

❦**What are they?** These extinct cephalopod molluscs lived in calcareous shells that, with few exceptions, were bilaterally symmetrical and spirally coiled (planispiral). They were mostly disc-shaped and divided into chambers. Ammonoids had well-developed eyes and numerous tentacles. They swam using water-jet propulsion, by expelling water to move forward.

❦**How big are they?** From a few centimetres up to a metre.

❦**How did they live?** These carnivorous predators and scavengers caught prey with their tentacles. They inhabited shallow seas – many could swim, but some larger forms were probably more restricted to the sea bed.

❦**How common are they?** The abundance of ammonoids in the fossil record, along with their rapid evolution and widespread distribution, makes them key to the subdivision of marine strata into biozones, or periods of time.

Amphibians – such as frogs, toads, newts and salamanders – class Amphibia, phylum Chordata

❦**Where are they found?** Continental water-laid deposits, especially fine river and lake muds.

❦**What age are they?** Devonian to the present day.

❦**What are they?** The vertebrate tetrapod animals that we recognize today as amphibians include the familiar frogs, toads, newts and salamanders, and the less familiar caecilians, which are legless. They are collectively known as the lissamphibians and are very different from their extinct Palaeozoic tetrapod ancestors. Among the most successful of the ancient extinct groups were the salamander-like temnospondyls (Carboniferous–Triassic), some of which were large, predatory, armoured forms and looked more like reptilian crocodiles, such as the 2-metre-long *Eryops* of early Permian times. Living frogs and toads are highly specialized but still primitive vertebrates in the sense that they require water into which their eggs and sperm are shed for external fertilization and development of tadpole-like larvae.

❦**How big are they?** From a few centimetres up to 3 metres long.

❦**How do they live?** Almost all amphibians require water for reproduction and some are entirely aquatic, but many live their adult lives on land. Most are predators.

❦**How common are they?** There are over 4000 living amphibian species, whose extinct ancestors had already diversified into 40 separate families by Carboniferous times.

Arthropods – phylum Arthropoda

The most abundant and diverse group of animals, the arthropods range from microscopic mites to giant spider crabs and extinct forms such as the eurypterids, which grew to 2 metres long. Important groups of fossil arthropods include the extinct trilobites and eurypterids, along with the living crustaceans and insects.

Asterozoa – starfish and brittle stars – phylum Echinodermata

🐚**Where are they found?** In a wide variety of marine deposits.

🐚**What age are they?** Ordovician to the present day.

🐚**What are they?** Like most echinoderms, the asterozoans have no obvious head or tail but a basic five-fold body symmetry with five or more arms around a disc-shaped body. The lower surface of the arms has rows of sucker-like tube feet that help the animal move slowly over the sediment and in some species burrow into soft sediment. The tube feet are also used for respiration and chemoreception (sensing certain chemicals in the environment).

🐚**How big are they?** From 1 to 25 cm in diameter.

🐚**How do they live?** Many starfish are carnivores that prey upon bivalves, whose shells they pull apart with their arms. The starfish then turns its stomach inside-out through its mouth, into the body space of the bivalve, where the digestive juices break down the bivalve tissues.

🐚**How common are they?** Starfish and brittle stars are not particularly common as fossils because their bodies tend to fall apart on death. The Asterozoa found as fossils tend to be those that have been buried whole.

Belemnites – class Cephalopoda, phylum Mollusca

🐚**Where are they found?** Mostly in marine shales and limestones.

🐚**What age are they?** Early Devonian/ Carboniferous to early Paleogene.

🐚**What are they?** Belemnites are the fossilized hard parts of squid-like cephalopods that had an internal shell-guard that is preserved as a bullet-shaped calcareous (calcite) fossil.

🐚**How big are they?** The fossils are up to

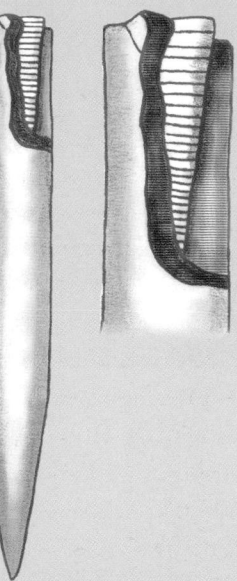

30 cm long, but the original animals were twice that.

🐚**How did they live?** They were active carnivorous predators, some of which swam in shoals, but were themselves preyed upon by larger marine vertebrates, such as the extinct ichthyosaurs.

🐚**How common are they?** Belemnites were particularly common during Jurassic and Cretaceous times. Sometimes large clusters of the bullet-shaped fossils are found. These may have been regurgitated by the large marine vertebrates that fed upon belemnites but could not defecate the sharp, pointed hard parts.

Birds – avialean dinosaurs – class Aves, phylum Cordata

🐚**Where are they found?** Because of their relatively fragile bones and their living habits, birds are not very well represented as fossils, except where their remains have been trapped and buried in fine-grained water-laid deposits such as those of lakes and lagoons.

🐚**What age are they?** Late Jurassic to the present day.

🐚**What are they?** Birds used to be regarded as a unique tetrapod group descended from the reptiles and characterized by the possession of feathers. However, since the discovery of feathered, flightless dinosaurs, it has been realized that feathers are not unique to birds. Birds also possess several anatomical features that link them with certain small, two-footed theropod dinosaurs. Birds are now regarded as a surviving group of avialean dinosaurs.

How big are they? From a few centimetres up to a 6-metre wingspan.

How do they live? Birds initially diversified in their habits in late Cretaceous times, with the evolution of shore birds and aquatic birds, but the main diversification occurred in Cenozoic times, following the extinction of the flying reptiles.

How common are they? Today, there are more than 9000 species of birds, belonging to about 153 families – considerably more than there are mammal species. There are at least 77 families that are now extinct.

Bivalves – clams – class Bivalvia, phylum Mollusca

Where are they found? Fossil bivalves are found in a wide range of marine and freshwater sedimentary rocks.

What age are they? Cambrian to the present day.

What are they? Compared with most molluscs, the bivalves have a highly modified form as the body is enclosed within two calcareous valves or shells (left and right) held together by an organic ligament and hinged by teeth and sockets. The shells are lined with flaps of mantle tissue that enclose the paired gills and the mouth, gut and muscular foot. The shells are opened by the springy ligament and closed by a pair of muscles. Movement is mostly produced

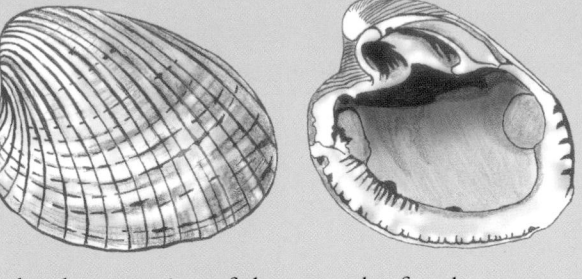

by the extension of the muscular foot between the valves into the surrounding sediment. However, a few bivalves also have adaptations for more dramatic means of escape, such as water-jet propulsion (scallops) and swimming (some *Limidae*, or file shells). Others, such as oysters, cement themselves to hard surfaces.

How big are they? From a few centimetres up to a metre.

How do they live? Most bivalves are filter-feeders, living on or in the sediment and using siphons to draw in water and any food particles, which are sieved out by the gills. The gills act as a breathing organ, extracting oxygen from the water. Some bivalves also feed on organic detritus, which they suck up from the sediment surface. Many bivalves burrow into the sediment to try to escape their numerous predators. However, some predators (such as certain carnivorous gastropods) can burrow after them.

How common are they? There are some 8000 living species of bivalves, and they are among the most common fossils. Many bivalves are preserved as moulds when the original shell carbonate was aragonite.

Brachiopods – lampshells – phylum Brachiopoda

Where are they found? Brachiopod fossils are found in a variety of marine sedimentary rocks. They are especially well preserved in limestones.

What age are they? Cambrian to the present day, but more common in the past.

What are they? These sea-bed-dwelling shellfish have a body enclosed in two calcareous dorsal and ventral (back and front) shells, held together with muscles and a hinge of teeth and sockets, that look superficially like molluscan bivalves. Normally, brachiopods have one shell bigger than the other and the bigger one is perforated with a hole through which a fleshy anchoring stalk projects. The name 'lampshell' refers to their resemblance to a Roman oil lamp.

How big are they? From a few millimetres up to 20 cm.

How do they live? These filter-feeders pump seawater through their interior shell cavity. They use a feathery structure (the lophophore) to sieve organic particles from the seawater.

How common are they? Some 4000 extinct genera are known, with only 350 living species.

Bryozoans – moss animals – phylum Bryozoa

Where are they found? Mostly in marine sediments, but some also live in fresh waters.

What age are they? Ordovician to the present day.

What are they? These aquatic animals live in colonies made of tens to thousands of millimetre-long individuals. Asexual reproduction occurs by budding from a single founder zooid, and individuals are interconnected.

The colony is protected by a calcareous skeleton of variable shape, ranging from three-dimensional forms to fan shapes and structures that encrust other surfaces.

How big are they? Colonies range from a few millimetres to about 20 cm.

How do they live? Filter-feeding bryozoan colonies were often important components of reef ecologies, especially during Palaeozoic times.

How common are they? The mineralized skeleton ensures a good fossil record throughout the Phanerozoic eon, although the delicate frameworks are often broken. Several groups of bryozoans became extinct at the end of the Triassic period, but they are still very abundant today. There are about 6000 species of living bryozoans, mostly living in the sea. Some 15,000 fossil species have been described.

Cephalopods – today, the 650 or so species of cephalopods include squids, cuttlefish, *Nautilus* and the octopuses, which all belong in the phylum Mollusca. The main fossil groups of cephalopods are the extinct ammonoids, belemnites and nautiloids.

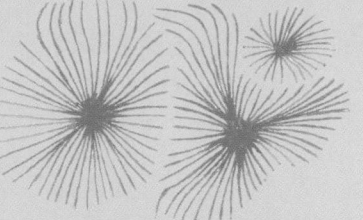

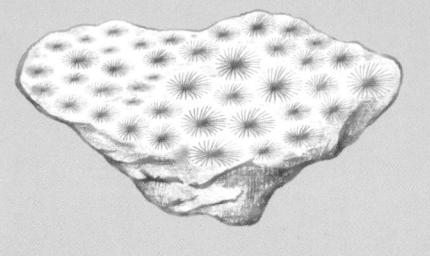

Cnidarians – corals, jellyfish and sea anemones. Of these, the corals are by far the most significant as fossils because of their abundance, the formation of massive reef structures and their importance to the evolution of marine ecosystems.

Conodonts – class Conodonta, phylum Chordata

☙**Where are they found?** Fossil remains are mostly found in marine limestones and shales.

☙**What age are they?** Cambrian to Triassic.

☙**What are they?** Conodonts are, as their name suggests, 'cone-shaped teeth' belonging to an extinct group of small eel-shaped animals whose biological affinities were a puzzle for 150 years. Their teeth varied in shape from simple cones to blades and flatter platform elements that were combined in pairs to form a feeding apparatus. The eventual discovery of the fossilized soft tissues of a conodont showed that these creatures were primitive vertebrates.

☙**How big are they?** Fossilized conodont teeth are mostly millimetre-sized (and rarely centimetre-sized). The animals were usually 1–5 cm long but could grow up to 50 cm.

☙**How did they live?** They were free-swimming marine predators that hunted small protistan, or single-celled, prey.

☙**How common are they?** Since the fossil teeth are made of calcium phosphate, they are often well preserved as fossils. Their fossil preservation, wide distribution and abundance in some sedimentary rocks, combined with their fast rates of evolution, have made them very useful for the subdivision of Palaeozoic marine strata.

Corals – class Anthozoa, phylum Cnidaria

☙**Where are they found?** Widespread in marine sedimentary rocks but most common in shallow-water reef limestones.

☙**What age are they?** Cambrian/Ordovician to the present day.

☙**What are they?** Biologically these are simple invertebrate animals like sea anemones with a radial symmetry (can be cut through the centre in two or more planes and produce halves that are mirror images) and no defined head or tail. They exist in both solitary and colonial forms. From the palaeontological point of view, corals, which secrete a calcareous (calcium carbonate) skeleton, are the most important cnidarians because they are well preserved as fossils. Fossil corals consist of the calcareous skeleton that is gradually secreted by the coral polyp, which lays down a series of layers and blade-like septa in particular patterns. Species can be recognized according to their distinctive form and pattern. The development of coral reefs has been important since Ordovician times, as they provide shelter and food for a great diversity of other organisms, from microbes and algae to shellfish and a wide variety of vertebrates. Reefs are the marine equivalent of the tropical rainforests, and modern reefs may support as many as a million other species.

Early Palaeozoic evolution of the corals produced two major groups, the tabulate and rugose corals, distinguished by details of their skeletons. The tabulates are all colonial and have horizontal partitions within a calcareous (calcitic) corallite skeleton. By contrast, the rugose corals are mostly solitary and have axial blade-like structures in the corallite with a superficially radial symmetry (in detail they are seen to be only bilaterally symmetrical, with two halves being mirror images). Both groups became extinct at the end of Permian times. The modern scleractinian corals evolved in Triassic times from a group of anemones that survived the extinction event. They have radially symmetrical, axial, blade-like septa that extend to the edge of the corallite, and their skeletons are made of a different calcium carbonate mineral (aragonite) than the calcite of the Palaeozoic corals.

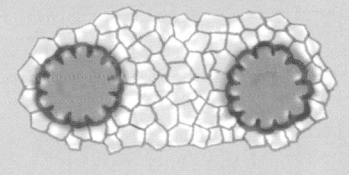

How big are they? From a centimetre to several metres. Coral reefs made of many colonial and solitary corals, plus other rock-forming organisms, are among the largest organic structures ever known, and can extend for hundreds of miles.

How do they live? The most abundant and well-known corals today are those of tropical reefs. These corals are dependent upon light because their tissues are inhabited by photosynthetic algae.

How common are they? Today, there are some 1300 species of corals, most of which live in shallow, warm, tropical seas. However, some corals can survive in deep, cold ocean waters.

Crinoids – sea-lilies – sub-phylum Crinozoa, phylum Echinodermata

Where are they found? Widespread in marine deposits but often best preserved in limestones, mostly as isolated plates.

What age are they? Mid-Cambrian to the present day.

What are they? This major group of echinoderms superficially look like plants and often have root structures and a long stem surmounted by a cup and feathery-looking branches. There are also stemless crinoids. Some of them can use their arms for swimming. The skeleton, like that of most echinoderms, is made of numerous calcareous plates held together by a thin coating of tissue. They have a five-fold symmetry with no distinct head or tail.

How big are they? From a centimetre to several metres high.

How do they live? Many live rooted to a hard surface – this is often the sea bed, but not always. Some are attached to other objects such as empty shells and floating logs. They are essentially filter-feeders, consuming organic particles in the seawater.

How common are they? Complete fossil crinoids are relatively rare since their skeletons disintegrate upon death, scattering their component parts into the surrounding sediment.

Crustaceans – with some 50,000 living species, the crustaceans are a large and very important group of arthropods with exoskeletons and jointed appendages. They range from well-known marine groups, such as the crabs, lobsters and shrimps (all decapods, with five pairs of limbs, ranging from Triassic times to the present day) and barnacles (Cretaceous to present), to the less familiar, small, bivalved ostracodes (Cambrian to present) and phyllocarids (Cambrian to present). Then there are the freshwater crayfish (Triassic to present) and isopods (Carboniferous to present), which include terrestrial forms commonly known as woodlice. Several of these groups, such as the crabs and ostracodes, have mineralized exoskeletons and as a result have good fossil records.

Dinosaurs – Dinosauria – class Sauropsida, phylum Chordata

Where are they found? As they were essentially terrestrial animals, dinosaur fossils are most commonly found in continental deposits. But many of the best-preserved dinosaur fossils are found in fine-grained water-laid deposits such as those of lakes, lagoons and near-shore, where dinosaur carcasses have been transported and buried.

What age are they? Mid-Triassic to end of the Cretaceous.

What are they? This group of reptiles is extinct, apart from their surviving avian descendants, the birds. The dinosaurs evolved from archosaur reptiles, and from late Triassic times they diversified rapidly to dominate terrestrial ecosystems throughout the Jurassic and Cretaceous periods until their extinction about 65 million years ago. There are two main groups of dinosaurs, based on the anatomical construction of the pelvis and hip bones. The bird-hipped ornithischians were all plant-eaters and mostly moved around on all four legs. They include groups such as the stegosaurs, ankylosaurs, ceratopians and bipedal ornithopods. By contrast, the lizard-hipped saurischians are subdivided into the plant-eating four-legged sauropods and the meat-eating bipedal theropods. It was from this latter group that the birds evolved in Jurassic times.

How big are they? From a few centimetres up to 25 metres long and 50 tonnes in weight.

How did they live? Both plant- and meat-eating dinosaurs varied enormously in their habits within terrestrial environments. The plant diet of the herbivores depended upon reach and tooth structure, with some long-necked sauropods able to stretch up several metres high into tree canopies while smaller forms browsed on low vegetation. The old idea that sauropods held their necks vertically has now been discounted, as the biomechanics of lifting blood up several metres to the brain would require enormous hearts that would not be very efficient. The meat-eaters also varied from scavengers and ambush-hunters such as *Tyrannosaurus rex* to more agile, fast-running pack-hunters such as the velociraptors.

How common are they? Over 600 genera have been described so far. Most are known from only a few specimens, but some genera are much better known, with tens and even hundreds of skeletons. By comparison with other common fossils, however, dinosaurs are exceptionally rare.

Echinoderms – as their name suggests, these

are animals with spiny skins. Their bodies have a basic five-fold symmetry and no distinct head or tail. Altogether they are a large and diverse group that includes the familiar sea urchins (Echinozoa – echinoids), starfish (Asterozoa) and crinoids (Crinozoa), and the less familiar blastozoans and homalozoans. Of these, the most important fossils are the sea urchins, starfish and crinoids.

Echinoids – sea urchins – class

Echinozoa, phylum Echinodermata

Where are they found? Widespread in marine deposits but best preserved in limestones, although even here they are mostly found as separate skeletal plates and spines.

What age are they? Ordovician to the present day.

What are they?
The sea urchins are marine invertebrates with globular to disc-shaped calcareous skeletons (known as 'tests') covered with protective spines. The body has a five-fold symmetry, seen in the rows of calcareous plates that make up its skeleton along with

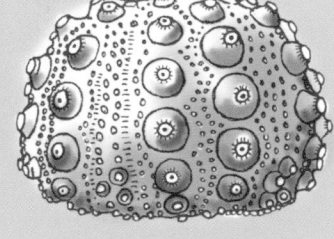

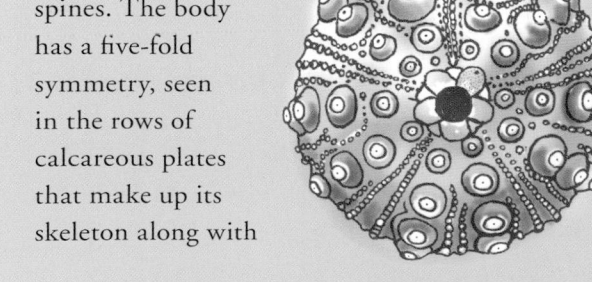

the rows of pores for the tube feet and tubercles for articulation of the spines. The skeleton is covered with a thin layer of living tissue, and muscles and ligaments to move both the body and the spines for protection from predators.

How big are they? From 1 cm up to 20 cm wide, including the spines.

How do they live? Sea urchins with spherical tests live on the sea bed. The shape of others is modified for burrowing.

How common are they? Complete fossils of sea urchins are relatively rare because the spines fall off after death and the test is easily broken. The most complete specimens are of burrowing forms that died within the sediment.

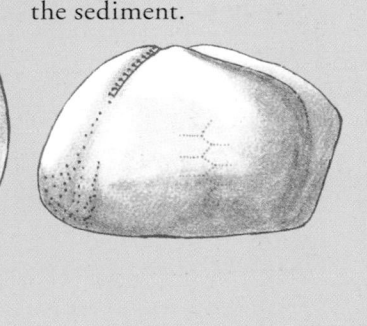

Eurypterids – class Eurypterida,

phylum Arthropoda

Where are they found? These extinct arthropods are found in a variety of deposits ranging from marine through brackish into freshwater and hypersaline.

What age are they? Ordovician to Permian.

What are they? With their elongated, dorsally flattened and segmented bodies and jointed appendages, these arthropods look superficially like scorpions. Some of them had large pincers.

How big are they? From a few centimetres to over 2 metres, these were some of the largest arthropods to have lived.

How did they live? Many were scavengers, but some were active predators.

How common are they? Being arthropods, eurypterids had to shed their exoskeleton in order to grow. Although the exoskeleton was not mineralized, pieces of the armour do survive as fossils. Whole animals are very rare as fossils and are mostly found in fine-grained, low-energy (where the body is less likely to be damaged by movement) mud deposits.

Fish – various taxa – jawless, jawed bony fish, jawed cartilaginous fish and lobe-finned fish

Where are they found? In a wide range of aquatic deposits, but often best preserved in quiet water sediments such as those of lagoons, lakes and certain inland seas.

What age are they? The jawless fish evolved in early Cambrian times and flourished in the lower Palaeozoic. Only a few survive today. The jawed fish, both bony and cartilaginous, first evolved in late Ordovician times. They diversified and still thrive today.

What are they? The earliest jawless fish of early Cambrian seas were the most primitive vertebrates from which all subsequent vertebrates evolved. The most important innovations included the evolution of jaws and teeth in the jawed fish (gnathostomes).

The jawed fish diversified into a number of groups in Devonian times, including the placoderms, which became extinct, the cartilaginous (chondrichthyan) fish that survive today and the bony (osteichthyan) fish.

The bony fish diverged into the ray-finned (actinopterygian) fish and the lobe-finned (sarcopterygian) fish, with paired muscular fins that eventually evolved into basic paired limb structures before the tetrapods emerged from the water in late Devonian times.

The ray-finned bony fish went through successive evolutionary changes, with a reduction in the heavy, bony scales to the thin overlapping scales of the modern Cenozoic bony fish (teleosts).

How big are they? From a few centimetres up to several metres (and exceptionally 20 metres).

How common are they? There are some 800 living species of cartilaginous fish today, compared with over 21,000 species of bony fish and just a few agnathans. In lower Palaeozoic times, the jawless fish were dominant forms in marine and fresh waters until they were out-competed by the jawed fish. The modern cartilaginous fish are largely restricted to marine waters, while the bony fish have also been able to exploit fresh waters. There are also several extinct groups of primitive fish, some of which were very successful at various points in Palaeozoic times.

Foraminiferans (forams) – sarcodines – phylum Foraminifera

Where are they found? Their fossils are mostly preserved in marine limestones.

What age are they? Cambrian to the present day.

What are they? These tiny-shelled unicellular (protistan) organisms are grouped with the more familiar naked forms such as the amoebas in the sarcodines. The shell is built by the protistan as a coiled series of chambers whose form varies from species to species and whose composition varies from calcareous to agglutinated masses of sand grains.

How big are they? Mostly millimetre-sized, but a few kinds, especially the nummulites, grew to a few centimetres (although rarely more than 6 cm) in diameter. The Egyptian pyramids are made of nummulitic limestones.

How do they live? Foraminiferans vary in their living habits from sea-bed-dwelling (benthic) forms to near-surface (planktonic) forms. Recovery of foraminiferan (or foram) shells laid down on the ocean floor over the last several millions of years has allowed the development of an important proxy measure of climate change. Analysis of the isotope ratio in the composition of their shells reflects the proportion of fresh and saline water at the time they were alive. Hence, it also measures the proportion of fresh water locked up in ice caps during Ice Ages.

How common are they? They may be incredibly abundant in nutrient-rich oceanic waters, so much so that their remains can accumulate on the ocean floor to form deep sea 'oozes' made almost entirely of foraminiferan shells. However, such 'oozes' are rarely preserved in the rock record because of plate tectonic processes. Nevertheless, in some limestones, fossil foraminiferans are very abundant and can be used as markers to subdivide the strata into time periods.

Gastropods – snails – class Gastropoda, phylum Mollusca

Where are they found? Gastropods are found in a wide variety of deposits from marine to freshwater and terrestrial.

What age are they? Cambrian to the present day.

What are they? Gastropods are among the most versatile of the molluscs in terms of their adaptation to differing environments. The body has a distinct head and tail separated by a flat, ventral creeping foot. The rest of the body is carried within a coiled, calcareous shell into which the animal can retreat for protection. Over geological time, gastropods have moved from their original marine habitat into freshwaters and then onto land.

There are about 15,000 living species of terrestrial snails. Some gastropods, such as the terrestrial slugs and marine sea-hares, no longer possess a shell and have little representation in the fossil record.

How big are they? From a few millimetres up to about 30 cm.

How do they live? Gastropods have acquired various lifestyles, from grazing on algae and plants to active carnivorous predation.

How common are they? Today there are about 40,000 species of snails. Gastropods are also among the most common fossils, found in a wide variety of rocks. They are often found in considerable abundance in environments where they had few competitors — marginal or extreme habitats ranging from brackish to hypersaline waters. However, since the calcareous shell is composed of the relatively unstable mineral aragonite, gastropod shells often dissolve away, leaving just moulds in the rock.

Graptolites – class Graptolithina, phylum Hemichordata

Where are they found? In a variety of marine sediments, but best preserved in shales and sometimes limestones.

What age are they? Mid-Cambrian to early Carboniferous.

What are they? The name 'graptolite' means 'writing in the rock'. Graptolites tend to look like strangely geometrical pencil lines or elongated bits of fret-saw blades with tiny serrations. A graptolite colony consists of series of interconnected independent bodies known as zooids that are budded asexually from a single founder zooid. The zooids construct a tubular organic skeleton of collagen-like proteins for protection. Each zooid has its own cup-shaped tube in which it lives and feeds. Graptolite fossils are the carbonized remains of the skeleton.

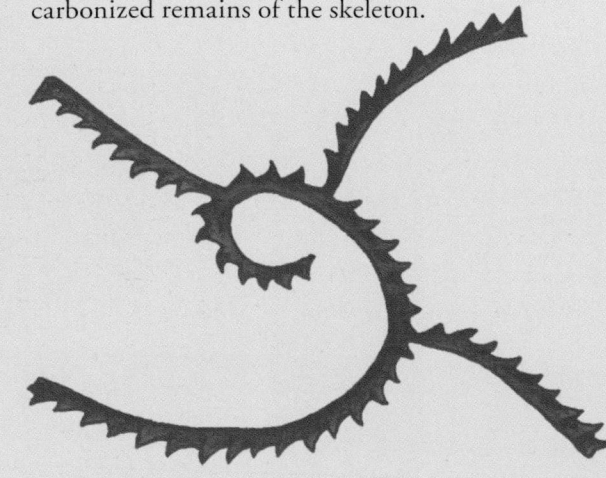

Studies of the closely related living pterobranchs suggest that, despite a superficially plant-like appearance, the graptolites, like the pterobranchs, were remarkably sophisticated colonial invertebrates close to the chordates.

How big are they? From a centimetre up to a metre long.

How did they live? The graptolites were all filter-feeders. There were two main groups: the dendroids, which mostly lived attached to the sea bed or some other hard surface, and the graptoloids, which were free-floating within the water column. They had limited powers of locomotion but may have had some buoyancy control that allowed them to adjust the depth at which they lived.

How common are they? Graptolites were often very abundant in lower Palaeozoic seas and oceans. Some shale rocks are known to contain masses of graptolite remains on every bedding plane. Their abundance, distribution,

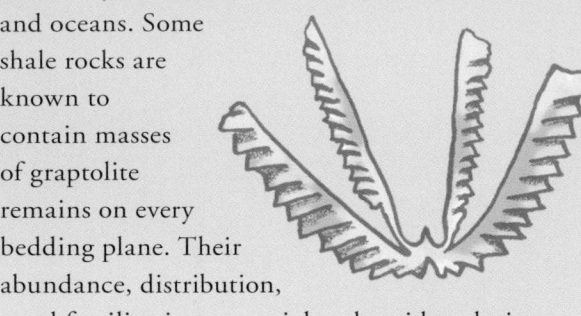

good fossilization potential and rapid evolution have made them excellent biozonal fossils for the subdivision of lower Palaeozoic strata.

Hominids – higher apes and humans – family Hominidae, order Primates

Where are they found? Being essentially terrestrial animals, fossils of higher apes and human ancestors are mostly found in terrestrial deposits, but they are very rare.

What age are they? The first higher ape fossils are found in early Miocene strata, about 10 million years old. The first human-related fossils are of late Miocene age, about 7 million years old.

What are they? This group of advanced primate mammals includes our human ancestors and *Homo sapiens*.

How big are they? From 1 to 2 metres tall.

How did they live? The higher apes mostly lived in wooded environments and were primarily plant-eaters, as were early humans, which lived in Africa. About 3 million years ago, some human-related species diversified their diet to include meat. As scavengers and hunters, they moved into more open woodland. By 2 million years ago, the first of these human relatives had expanded beyond Africa into the Middle East and Asia. About 85,000 years ago, *Homo sapiens* expanded out of Africa.

How common are they? There are about 20 human-related species, all extinct except for *Homo sapiens*. Overall, fossils are very rare. Many species are only known from a handful of bones, and just a few are known from complete skeletons.

Insects — class Insecta, phylum Arthropoda

Where are they found? Mostly in the fine-grained deposits of estuaries and lakes and occasionally near-shore marine deposits. Also exceptionally well preserved in amber resins.

What age are they? Silurian to the present day.

What are they? Insects are a group of arthropods in which the basic body plan is divided into three parts: a head with six segments, a body with three segments and an abdomen with eleven or fewer segments. The head has compound eyes, chewing mandibles and sensory antennae, while the body has two pairs of wings and each segment carries a pair of walking legs. However, there are many departures from this plan and many insects are highly specialized.

How big are they? From a few millimetres to around 70 cm.

How do they live? Thanks to the early evolution of flight within the insects, this is an exceptionally diverse and widely distributed group of organisms within terrestrial environments including fresh water.

How common are they? Insects are among the most common animals ever to have existed, with around a million living species (of which 300,000 are species of beetle). Because insect exoskeletons are not mineralized, their fossil representation is not great. However, in some exceptional circumstances, such as amber, very large numbers are preserved, and some 40,000 fossil species have been described.

Mammals — class Mammalia, phylum Chordata

Where are they found? Widespread throughout terrestrial and aquatic deposits (though generally never common as fossils except in some exceptional circumstances). Most mammal fossils consist of bony fragments and teeth.

What age are they? The earliest true mammals are late Triassic in age to the present day.

What are they? Biologically, mammals are defined by a number of characteristics such as being warm-blooded and hairy, bearing live young, and having milk-secreting mammary glands. None of these factors is normally preserved in fossils. However, there are some anatomical features that evolved towards the mammalian condition, especially related to the development of the jaw as a single dentary bone with the associated bones making up the tiny bones of the middle ear.

How big are they? From centimetre-sized pigmy shrews to 25-metre-long blue whales.

How do they live? The diversification and expansion of the mammals since Jurassic times has resulted in them occupying a remarkable diversity of environments and ecological niches, from the deep oceans to polar wastes and the air. The post-Cretaceous expansion of the mammals produced a large number of groups in early Cenozoic times that are now extinct, such as the creodonts.

How common are they? There are some 5400 species of mammal today, divided into some 20 major groups.

Molluscs – phylum Mollusca

The molluscs are one of the most important, abundant and diverse groups of fossils, with some 60,000 fossil species and 50,000 living species, most of which are clams (bivalves) and snails (gastropods). The mollusc body form is very variable, ranging from the free-swimming squids to the benthic clams enclosed in their bivalved shells. The phylum also includes the cephalopods (such as living squids, octopus and extinct ammonites) and minor living groups, such as the limpet-like monoplacophorans (Cambrian to present), chitons (polyplacophorans, Cambrian to present), tusk-shells (scaphopods, Ordovician to present) and the extinct snail-like bellerophontids (Cambrian to Permian).

Nautiloids – class Cephalopoda, phylum Mollusca

Where are they found? In a wide variety of marine sediments, but they are best preserved in limestones.

What age are they? Late Cambrian to the present day. The pearly *Nautilus* is the only externally shelled cephalopod to survive today.

What are they? These marine molluscs include the most ancient cephalopods known and mainly have simple, straight, conical, chambered shells made of calcium carbonate (aragonite). In early Mesozoic times, bilaterally symmetrical, spirally coiled (planispiral) forms evolved. An interconnecting tube runs through the chambers, which are filled with a liquid/gas mixture; the relative proportions of this mixture can be altered by the animal to control its buoyancy. The walls between the chambers are simple by comparison with those of later, more evolved ammonoids.

How big are they? From a few centimetres up to a metre (but exceptionally 10 metres long).

How did they live? Most were free-swimming, marine carnivorous predators that lived a bit like modern squid, but some of the large forms, with relatively large shells, lived on the sea bed.

How common are they? Around 1000 fossil genera are known. They were often very abundant in Palaeozoic seas, with vast accumulations of their shells sometimes forming *Orthoceras* limestones.

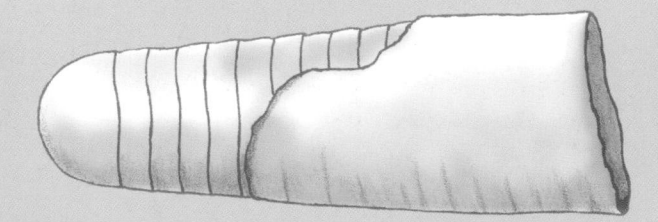

Plants

Where are they found? Fossil plants are very widespread throughout different strata, but most plants are terrestrial, so their fossils are found in continental deposits, especially swamp and bog environments where their remains may accumulate in sufficient abundance to form peats. With burial and progressive compression, plants can be transformed into lignites, coals and anthracites. Plant debris may also be carried by rivers and deposited in near-shore environments. Many specialized plants also live in marsh and delta environments, where their remains are commonly preserved. Land plants are mostly found in aquatic depositional environments, where their delicate tissues stand a better chance of preservation.

What age are they? Fossil spores suggest that some vascular plants (with tissues that conduct water, minerals and photosynthetic products through the plant) may have evolved in late Ordovician times, but fossils of adult vascular and upright-growing land plants first appeared in early Silurian times and remain to the present day.

What are they? Plants use light energy from the Sun and photosynthesis to form the base of the terrestrial food chain. They have transformed land surfaces from barren, rocky wastes to soil-covered environments that support a huge diversity of vegetation around the world. From centimetre-sized Silurian leafless organisms growing in watery environments, plants first diversified and grew to tree size in late Devonian times.

The first extensive forests formed in Carboniferous times and were populated by groups of plants that are largely unfamiliar today, such as the clubmosses (lycopsids, late Silurian to present), horsetails (sphenopsids, late Devonian to present), seed ferns (pteridosperms, Carboniferous to present) and true ferns (late Devonian to present).

By Mesozoic times, the dominant plant groups had changed to cycads (Permian to present), bennettitaleans (mid-Triassic to late Cretaceous) and particularly the surviving conifers (Carboniferous to present). By Cretaceous times the flowering plants had appeared and they dominated land environments in Cenozoic times as they co-evolved with their insect, bird and mammal pollinators.

How big are they? From a few centimetres to 50 metres high.

How do they live? Plants not only require sunlight for photosynthesis but also water and nutrients from the surrounding environment, usually obtained from soil through their root structures. Initially, land plants were restricted to living in watery environments for reproduction. Considerable development of the reproductive system was required before plants could colonize drier and more upland environments in Triassic times. With their long evolutionary history, plants have adapted to a wide variety of living conditions, from swamps, hot deserts and frigid polar regions to mountain-tops and shallow seas.

How common are they? Today, there are about 250,000 species of flowering plants (angiosperms), including 10,000 species of grasses. The number of fossil plants is difficult to estimate because different parts of the plants have different names. A whole plant is rarely preserved in the fossil state.

Reptiles – class Reptilia, phylum Chordata

Where are they found? The reptiles have diversified to occupy a very wide range of environments and ecological niches, from the deep oceans to tropical deserts and the air. Their fossil remains are found in many different depositional environments, from marine to terrestrial. Most fossils tend to be isolated bones and teeth – complete skeletons are rare and are mostly found in low-energy environments (where the bones are less likely to be damaged by movement) such as lagoons, lakes and swamps.

What age are they? Carboniferous to the present day.

What are they? Essentially, reptiles are tetrapods that have evolved an independence of water and the ability to reproduce on land through internal fertilization and egg-laying. The fertilized embryo is covered with a porous membrane – the amnion – and subsequently a shell before being laid by the mother. Collectively known as the amniotes, the group includes the birds and mammals, which have evolved from the reptiles.

Reptiles vary from cold- to warm-blooded, and very large forms have such a large body volume to skin area that they can retain body heat. There are many extinct reptile groups, ranging from the marine ichthyosaurs and mosasaurs to the flying pterosaurs and the land-living dinosaurs.

How big are they? From a few centimetres to 25-metre-long dinosaurs.

How do they live? As tetrapods, the reptiles evolved from water-living animals and exploited a wide range of terrestrial environments, but some took to the air and others returned to the water in early Triassic times. To begin with, the reptiles were meat-eaters and ranged from small insect-eaters and fish-eaters to larger carnivorous predators, but plant-eaters soon evolved. This pattern of carnivores giving rise to herbivores was repeated many times as different reptile groups evolved.

How common are they? Today, there are some 16,800 species of reptiles, of which some 6800 are snakes and lizards. Living birds, now considered avialean dinosaurs, number 9600 species.

Sponges – phylum Porifera

Where are they found? Widespread in marine deposits from deep ocean to shallow waters and also fresh waters.

What age are they? From late Precambrian to the present day.

What are they? Sponges are groups of cells that cooperate to form a multicellular colony with very varied shapes, from globular and vase-shaped to branching forms. They have highly porous walls supported by a variety of skeletal materials ranging from silica (glass sponges) and carbonate (calcareous or stony sponges) to protein (bath sponges). Some have skeletons of mixed composition (demosponges).

How big are they? From a few millimetres up to a metre.

How do they live? Sponges are filter-feeders that grow attached to a sediment surface or other substrate such as another organism. They sieve organic particles from the water.

How common are they? Today, there are some 10,000 or more living species. At times in the past, they have been abundant enough to form reefs.

Tetrapods – all vertebrates with four limbs are tetrapods, so this group includes the amphibians, reptiles, mammals (including humans) and birds. The limbs are highly modified in many members of these groups to aid activities such as swimming and flying. In some groups, such as the snakes, caecilians (an order of limbless amphibians) and limbless lizards, the limbs have been lost altogether through evolution. The first tetrapods evolved from the sarcopterygian fish, which had paired muscular fins.

Trilobites – class Trilobita, phylum Arthropoda

Where are they found? Trilobites are found in many marine deposits but tend to be best preserved in limestones. The paired appendages are not normally preserved.

What age are they? From early Cambrian to late Permian.

What are they? Trilobites are an extinct group of arthropods. Like many arthropods, they are covered with a tough organic exoskeleton that is also mineralized in trilobites. The body is typically oval in plan and flattened, with a distinctive head carrying compound eyes (similar to those of flies). The thorax, divided into articulated segments, separates the head and tail shield. In times of danger, the tail could be rolled under the head to protect the vulnerable underside and limbs. The limbs and other appendages are jointed for movement. For growth, the exoskeleton had to be shed periodically. This moulting process is known as ecdysis. Most trilobite fossils are actually pieces of moulted exoskeleton.

How big are they? Trilobites range from a few millimetres up to 72 cm long.

How did they live? Most trilobites lived on the sea bed, scavenging for organic detritus. Some burrowed into the sediment and others could swim.

How common are they? Like so many groups of arthropods, the trilobites evolved into many different species. Today, some 15,000 fossil species have been described. It is certain that, in the coming years, many more will be discovered.

Many of these fossil species can be shown to form evolving and diverging lineages. In fact, trilobites provide one of the best examples of speciation over time known from the fossil record. As a result they can also be used to subdivide sequences of strata into trilobite biozones, or time periods.

GAZETTEER

RESOURCES

- **British Geological Survey (BGS)**
  www.bgs.ac.uk
- **Countryside Council for Wales**
  www.ccw.gov.uk
- **Dinosaur Coast programme**
  www.dinocoast.org.uk
- **Dinosauria Online**
  www.dinosauria.com
- **Discovering Fossils**
  www.discoveringfossils.co.uk
- **Dorset Coast Forum**
  www.dorset-cc.gov.uk/dcf
- **The Geological Society of London**
  www.geolsoc.org.uk
- **The Geologists' Association**
  www.geologists.org.uk
- **The Joint Nature Conservation Committee (JNCC)**
  www.jncc.gov.uk
- **Jurassic Coastline**
  www.jurassiccoastline.com
- **National Trust**
  www.nationaltrust.org
- **National Trust for Scotland**
  www.nts.org.uk
- **Natural England**
  www.naturalengland.org.uk
- **Open University Fossil Detectives website**
  www.open2.net/fossildetectives
- **The Palaeontological Association (PalAss)**
  www.palass.org
- **Rockwatch**
  www.rockwatch.org.uk
- **Scottish Natural Heritage**
  www.snh.org.uk
- **The Society of Vertebrate Palaeontology**
  www.vertpaleo.org
- **UK Fossils Network**
  www.ukfossils.co.uk

- **Walking with Dinosaurs**
  www.bbc.co.uk/dinosaurs
- **American Museum of Natural History (AMNH)**
  www.amnh.org
- **National Geographic**
  www.nationalgeographic.com
- **Smithsonian Institution**
  www.si.edu
- **United States Geological Survey (USGS)**
  www.usgs.gov
- **University of California, Museum of Palaeontology**
  www.ucmp.berkeley.edu
- **The Association of UK RIGS**
  www.ukrigs.org.uk

Advanced further reading
- **Benton, M.J.,** *Vertebrate Palaeontology*, 3rd edition (Blackwell Publishing, Oxford), 2005
- **Brenchley, P.J. and Rawson (eds),** *The Geology of England and Wales*, 2nd edition (The Geological Society, London), 2006
- **Clarkson, E.N.K.,** *Invertebrate Palaeontology and Evolution*, 4th edition (Blackwell Science, Oxford), 1998
- **The Natural History Museum,** *British Palaeozoic Fossils* (Intercept Ltd, Andover), 2002;
  British Mesozoic Fossils, (Intercept Ltd Andover), 2001;
  British Cenozoic Fossils (Intercept Ltd, Andover), 2001
- **Trewin, N.H.,** *The Geology of Scotland*, 4th edition (The Geological Society, London), 2002

Overview

Much of the enjoyment of being a fossil detective is in finding the clues to past life and knowing where to look for them. Fossils are found in rock strata that outcrop here and there in the landscape. To get a real appreciation and feel for fossils requires getting out and about and looking for them yourself. For those living in countryside where there are fossiliferous rocks, it is usually easy enough to make local enquiries about where to find fossils, but urban dwellers face more of a challenge.

Access to the countryside is often restricted by private land ownership, and most of our famous fossil sites are designated Sites of Special Scientific Interest (SSSIs), which are legally protected. The easiest free access to fossiliferous strata is along our beautiful and geologically varied coastline. But remember that coasts often have cliffs and can be dangerous. For the novice to learn where to go to freely collect fossils in safety, it is best to contact your local group of geological enthusiasts and ask them to guide you.

Local libraries and museums are good places to start in your search for information, as are web-based searches through national geological organizations. This gazetteer provides details on a selection of indoor and outdoor attractions and locations throughout Britain. These can supply you with information and advice to help get you started.

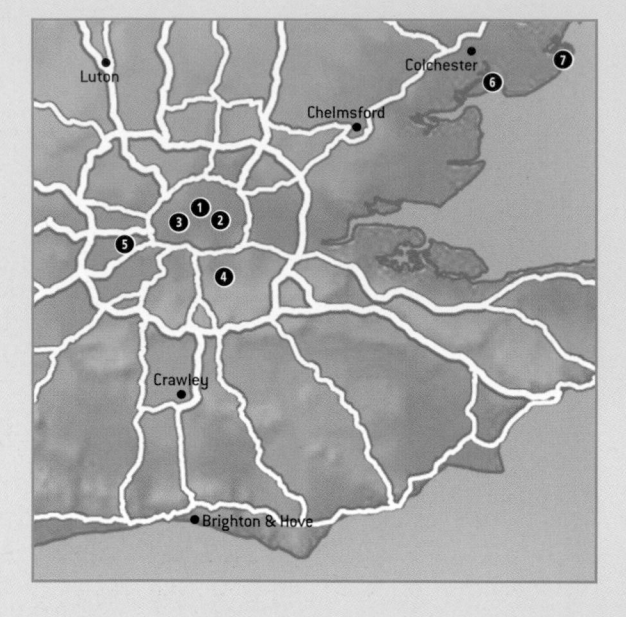

London

INDOORS

Hunterian Museum ❶
The Royal College of Surgeons of England
35–43 Lincoln's Inn Fields, London WC2A 3PE
Tel: +44 (0)20 7869 6560
Fax: +44 (0)20 7869 6564
E: museums@rcseng.ac.uk
www.rcseng.ac.uk/museums

An introductory gallery at the entrance leads you into the
museum and its collections, which include many thousands of
natural history specimens prepared during the nineteenth and
twentieth centuries, and a large palaeontology collection. The
Crystal Gallery has more than 3000 specimens, including species
of plants and animals collected by the eighteenth-century surgeon
and anatomist John Hunter. These include some rare specimens
such as the kangaroos brought back by Sir Joseph Banks from
James Cook's voyage of 1768–71, and a selection of specimens for
the royal collection at Kew, which were prepared by Hunter for King
George III and Queen Charlotte. These specimens are housed in eight
spectacular showcases enclosing the museum's central atrium. The
surrounding exhibitions explore the social and scientific context of
Hunter's life and work. A dedicated computer terminal within the
gallery provides access to detailed scientific and historical
information about the specimens. A virtual museum tour, viewing all
the galleries, is available via the website. The museum offers free
weekly guided tours and curator-led group visits.

Museum of London ❷
150 London Wall
London EC2Y 5HN
Tel: 0870 444 3851
Fax: 0870 444 3853
E: info@museumoflondon.org.uk
www.museumoflondon.org.uk

With a collection of more than a million objects, the Museum
of London is the world's largest urban history museum. Of
particular interest to fossil detectives is the London Before
London gallery, which explores the story of the Thames Valley
and the people who lived there from 450,000 years ago to the
Roman invasion and founding of Londinium in 50 CE. In the centre
of the gallery, a spectacular 'River Wall' features more than
300 objects dredged from the depths of the Thames – many of
them swords laid there as sacrifices to the gods. The gallery also
contains the remains of one of the oldest people found in the
region. The skeleton is more than 5000 years old and is displayed
alongside a facial reconstruction.

The entire exhibited collection can be browsed online, and you can
also play games, see the layout of the gallery and more on the
extensive website.

Natural History Museum ❸
Cromwell Road
London SW7 5BD
Tel: +44 (0)20 7942 5000 (general enquiry line)
www.nhm.ac.uk

This is the largest and most important natural history collection in
the world, with more than 70 million specimens. Among them are
some 9 million fossils, including one of only six specimens of
Archaeopteryx, the earliest known flying bird. Discover the stories
that fossils can reveal, including the hunting habits of early humans,
and the types of oversized trees that used to dominate Britain.
The collection includes the world's oldest known fossil insect,
information on fossil folklore, a dinosaur directory of fossils found in
Britain, virtual specimens, fossil plants of Britain, the giant sloth
Megatherium and other giant 3D sculptures.

The Central Hall contains two collection highlights: a *Diplodocus*
skeleton and 1300-year-old giant sequoia. In the dinosaur gallery,
explore the 160 million years of the dinosaur era: what dinosaurs
looked like, what they ate, how they evolved and why they died out.
Inspect the *Triceratops* skeleton and see the animatronic *T. rex*.

Other galleries of interest to fossil detectives include: the Fossils
from Britain gallery, which shows how fossils are preserved and,
over time, are transformed into sedimentary rock; the Fossil Marine
Reptiles gallery, located on Waterhouse Way, with some of the

finest examples of marine reptile fossils found in Britain; and the From the Beginning gallery, which takes you back to the Big Bang and through the history of time itself to explore the distant past, discover early sea creatures and mammals, and take a peek into the Earth's future. The museum's website includes an Earth Lab datasite of UK fossils, minerals and rocks.

OUTDOORS

Crystal Palace Park ❹
Thicket Road
London SE20 8DT
Tel: +44 (0)20 8778 9496 (park information line)
www.crystalpalacepark.org

An information centre, including the park rangers' office, is located near the Penge Gate entrance, opposite the park café. This historic park features 27 Grade 1 listed life-size prehistoric monster sculptures (including the earliest known dinosaurs), which were made in the nineteenth century. They are located on and around the islands of the Lower Lake. The models were restored in 2003 to their original colours and reconstructed to match the original designs. The park's team of rangers provides a leaflet about the Prehistoric Monster Trail and offers guided tours of the park and of the dinosaurs. There are also slideshow presentations on the history of the park and the dinosaurs. The tours can be specifically catered to groups' needs (including age and mobility access).

Royal Botanic Gardens ❺
Kew, Richmond
Surrey TW9 3AB
Tel: +44 (0)20 8332 5655 (24-hour information line)
E: info@kew.org
www.kew.org

The Royal Botanic Gardens are situated near Richmond, about 6 miles southwest of London. They are set over a large area of 300 acres, so a little time-planning will help you see the highlights that interest you. For fossil detectives, Evolution House is a fascinating walk-through experience of plant evolution over 3.5 billion years. An information leaflet about Evolution House is available inside the glasshouse. The landscape inside the entrance shows what the surface of the Earth may have looked like shortly after its formation. See how plant forms have changed over hundreds of millions of years, from the tiny first colonizers through mosses, ferns and giant clubmosses to the conifers and flowering plants that predominate today. Three major periods illustrate the long history of plant evolution: the Silurian, Carboniferous and Cretaceous. A volcano, glowing red lava and dinosaur footprints take you through the history of evolution. The display encompasses stromatolites, *Cooksonia* (the first plants that adapted to life on land), liverworts, mosses, selaginellas and the first ferns.

Guided Geology Walks
For details of guided walks on London's building stones and urban geology, as well as other fossil excursions in the London area, contact:

The Geologists' Association
Burlington House
Piccadilly
London W1J 0DU
Tel: +44 (0)20 7434 9298
Fax: +44 (0)20 7287 0280
E: geol.assoc@btinternet.com
www.geologists.org.uk

THAMES AND THAMES ESTUARY: FOSSIL HOTSPOTS

London
Unprecedented building and excavation in the late nineteenth century unearthed many fossil remains around the capital. The fossil evidence found so far includes hippopotamus and elephant remains beneath Trafalgar Square, woolly mammoth fossils under the Strand, reindeer fossils at South Kensington station and woolly rhinoceros remains under Battersea Power Station. The complex series of gravels, sands, silts and clays contains abundant fossil remains of plants and animals, and, most interestingly, large, extinct mammals. Many of these fossils are held at the Natural History Museum in South Kensington.

Essex
Colne Estuary: A large estuary that runs southeast of Colchester. One site of particular geological interest is the Cudmore Grove Country Park.

Cudmore Grove Country Park ❻
Bromans Lane
East Mersea
Essex CO5 8UE
www.essexcc.gov.uk

The park is located at the eastern end of Mersea Island at East Mersea, 9 miles south of Colchester. It can be found at the end of Bromans Lane and is well signposted. There are areas of particular fossil interest here in the low cliffs and foreshore.

Walton-on-the-Naze: ❼ Clay fossils, particularly sharks' teeth, turn up all around the Essex coast, but the most famous site is at Walton-on-the-Naze, where the beach is popular with collectors. The London Clay here – along with Red Crag – is one of the reasons Walton is a site of international importance. The clay has yielded superbly preserved bird fossils, including several species of parrot.

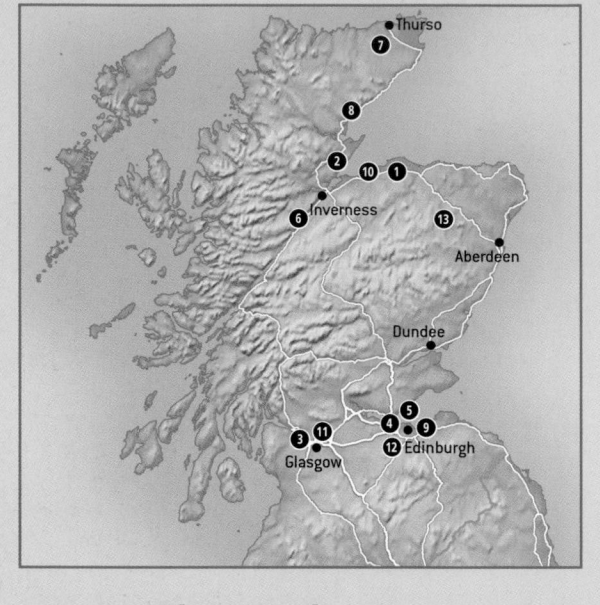

Scotland

INDOORS

Elgin Museum ❶

1 High Street, Elgin
Moray IV30 1EQ
Tel/Fax: +44 (0)1343 543675
E: curator@elginmuseum.org.uk
www.elginmuseum.org.uk

Situated at the eastern end of High Street, this Grade 1 listed
independent museum was built in 1843 from local fossiliferous
sandstone. There is an excellent permanent display of local fossil
fish and reptiles from the Permian and Triassic periods, including
Scotland's earliest known dinosaur. Some of the slabs found at
nearby Clashach Quarry, imprinted 250 million years ago with
reptile tracks and arthropod traces, are also on display. The
museum has permanent displays of local history and
archaeology, including a spectacular assembly of Pictish stones,
a new Victoriana display and a gallery of changing exhibitions.

Hugh Miller House and Cottage ❷

Church Street, Cromarty IV11 8XA
Tel: +44 (0)1381 600245 (general information)

One of the Highlands' best-preserved towns, Cromarty is about
22 miles northeast of Inverness. It is the birthplace of the
Scottish geologist Hugh Miller (1802–56), whose thatched
cottage is now a museum devoted to his life and work. It has a

display of fossils on permanent loan from the National Museums of
Scotland Miller collections. As a boy Miller excavated with his great-
grandfather's hammer, and his abiding interest in rocks and fossils
was fostered by the surrounding coast and hills. Many of his
collection of 6000 fossils are in the National Museums of Scotland.

Hunterian Museum ❸

Gilbert-Scott Building
University Avenue, University of Glasgow
Glasgow G12 8QQ
Tel: +44 (0)141 330 4221
Fax: +44 (0)141 330 3617
E: hunter@museum.gla.ac.uk
www.hunterian.gla.ac.uk

The Hunterian is spread across four sites on the University of
Glasgow campus, which is located in the Hillhead district,
2 miles west of the city centre. The bulk of the fossil material is
displayed in the main university building on University Avenue. Key
exhibits are the plesiosaur *Cryptoclidus*, several large ichthyosaurs,
the world's most complete reconstruction of the hominid 'Lucy' and
the remains of the unusually large fish *Leedsichthys*.

The National Museum of Scotland ❹

Chambers Street, Edinburgh EH1 1JF
Tel: +44 (0)131 225 7534
www.nms.ac.uk

Situated in the old town, a few minutes' walk from the Royal Mile, this
is one of the six National Museums of Scotland. The museum's
comprehensive natural-science collections focus on zoology and
geology; botany is represented by fossil plants. Collections of
several million specimens represent the astonishing diversity of
Scotland's geological background, including fossils showing
evidence of the earliest forms of life. The collection also covers
rocks and geological structures, birds and mammals, amphibians
and molluscs, insects and spiders, and marine invertebrates.

Our Dynamic Earth, Edinburgh ❺

112–116 Holyrood Road
Edinburgh EH8 8AS
Tel: +44 (0)131 550 7800
Fax: +44 (0)131 550 7801
E: enquiries@dynamicearth.co.uk
www.dynamicearth.co.uk

This science and natural history attraction is both educational and fun.
Set in a striking tent-like structure based in a former brewery in the
heart of Edinburgh it offers lots of fascinating Earth facts. Inside,
there are 11 galleries featuring multi-sensory insights with video
presentations and hands-on installations. Each gallery investigates
themes such as the immensity of geological time compared to the

human period. In the Casualties and Survivors gallery you can follow the path of evolution and come face to face with some of evolution's winners and losers – discover how mass extinctions occurred and what the dinosaurs may have looked like if the meteorite hadn't struck.

The Loch Ness Exhibition Centre ❻

Drumnadrochit, Loch Ness
Inverness-shire IV63 6TU
Tel: +44 (0) 1456 450573

Discover the famous story of the monster at this visitor attraction on the banks of Loch Ness. Keeping the legend centre stage, the attraction explores Scotland's geological past preserved in the Loch sediments. Information on loch cruises is also available.

OUTDOORS

Achanarras Quarry ❼

Achanarras, near Spittal
North Highland

Achanarras Hill is 2 miles south of Halkirk. Turn west off the A9 at Mybster crossroads onto the B870 towards Westerdale. After approximately half a mile, turn right onto a track beside a plantation that leads to a car park. A short walk up Achanarras Hill from the car park takes you to the former quarry. The site is one of the most famous fossil-fish sites in Britain and a designated National Nature Reserve with an area of 43 ha. It is an SSSI of international significance, owned by Scottish Natural Heritage (SNH). A new shelter with a geological timeline is on site. Anyone may fossil-hunt in the spoil heaps provided they follow the Scottish Fossil Code. Further information can be obtained from Scottish Natural Heritage.

Scottish Natural Heritage ❽

Main Street, Golspie
Sutherland KW10 6TG
Tel: +44 (0)1408 633602
Fax: +44 (0)1408 633071
www.snh.org.uk

Arthur's Seat and Salisbury Crags ❾

Holyrood Park, Edinburgh

Holyrood Park (Queen's Park) is a 265-ha wilderness of mountains, crags, lochs, moorland, marshes, fields and glens – all within walking distance of Edinburgh city centre. Its main feature is Arthur's Seat, the igneous core of an extinct volcano and the highest of Edinburgh's hills. Another dominating feature of the Edinburgh skyline is the adjoining precipitous Salisbury Crags lying directly opposite the gates of Holyrood House. Observations of the formations of Arthur's Seat and Salisbury Crags greatly influenced James Hutton's geological theories.

Clashach Quarry ❿

Hopeman, near Elgin
Morayshire

Clashach is an active quarry that works the Permian sandstone of the coastal cliff, 1 mile east of Hopeman. It is an SSSI, with some 300 trackways and arthropod traces. A selection with interpretative panels is on display next to the coastal path.

Fossil Grove ⓫

Victoria Park, Victoria Park Drive North
Glasgow G14 1BN
Tel: +44 (0)141 950 1448
www.glasgowmuseums.com

Fossil Grove is situated in Victoria Park in Scotstoun, in the west of Glasgow. The fossilized tree stumps on view were discovered in 1887 when an old quarry was being landscaped during the creation of the park. What you see is a small corner of a vast ancient forest, about 330 million years old, preserved in stone. The most striking feature is the 11 fossil tree stumps preserved in the position in which they once grew. They are the remains of an extinct type of plant known as a giant clubmoss.

The James Hutton Memorial Garden ⓬

St John's Hill, Edinburgh
www.james-hutton.org

St John's Hill is the site of the house and garden in which James Hutton lived. A visitor once wrote, 'His study is so full of fossils and chemical apparatus of various kinds that there is barely room to sit down.' The garden features a display that illustrates the main themes of Hutton's remarkable geological work. The website features The James Hutton Trail. The garden is best approached from Holyrood Road; 100 metres east of the junction with the Pleasance, turn into Viewcraig Gardens and walk up past the entrance to the car park. About 50 metres from Holyrood Road, a flight of steps leads up to the garden.

Rhynie Devonian Site ⓭

Rhynie, Aberdeenshire

Rhynie village is about 30 miles northwest of Aberdeen, and is famous as a key palaeobotanical site. Rhynie has provided some of the best fossil specimens of the primitive *Rhynia*, one of the earliest land plants, dating from the Devonian period. The Rhynie Chert is a site of exceptional fossil preservation; there is no outcrop of the chert, only loose blocks in a field and incorporated into stone walls. Being generally microscopic, Rhynie plants and animals make poor fossils for display purposes, but there are good online exhibitions at www.abdn.ac.uk/rhynie and www.uni-muenster.de/geopalaeontologie/palaeo/palbot/erhynie.html.

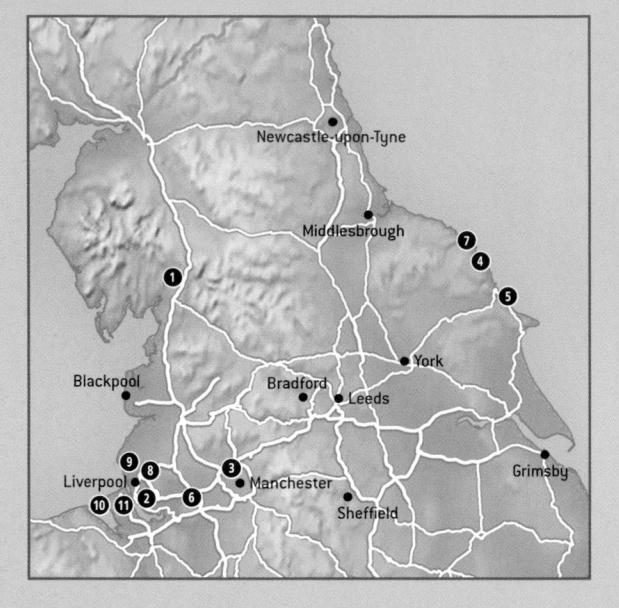

The North of England

INDOORS

Kendal Museum ❶

Station Road, Kendal, Cumbria LA9 6BT
Tel: +44 (0)1539 721374
Fax: +44 (0)1539 737976
E: info@kendalmuseum.org.uk
www.kendalmuseum.org.uk

The Lake District Natural History Gallery on the first floor starts with a geology display. Collections include a wealth of fossils, local shales, flags, grits and slates, and many local minerals and rock types – in all, a total of some 1000 fossils and 400 rocks. Among the vertebrate material there are a few ichthyosaur specimens and a fine fossil fish, *Dapedius*. The museum offers a regular programme of fossil-rubbing and other children's events.

Liverpool John Lennon Airport (JLA): Fossil Mystery Tour ❷

Liverpool L24 1YD
www.liverpoolairport.com

JLA, in partnership with the Liverpool Geological Society, has launched a Fossil Mystery Tour for visitors to explore the creatures that shared their footsteps more than 250 million years ago. The guide highlights the main types of fossils that can be seen in the limestone floor slabs of the terminal building and also offers a Fossil Mystery Tour around the terminal itself.

The Manchester Museum ❸

University of Manchester, Oxford Road
Manchester M13 9PL
Tel: +44 (0)161 275 2634
Fax: +44 (0)161 275 2676
www.manchester.ac.uk/museum

The museum is located on the south side of Manchester city centre with a visitor centre on Oxford Road. The Prehistoric Life gallery features some of the best of the museum's 250,000 fossils, including an impressive range of fossil fish and plants and a massive slab that shows the hand-like tracks of *Chirotherium*. Other highlights are marine reptiles, including plesiosaurs and ichthyosaurs found on the Yorkshire coast, and Stan, a cast of a *T. rex*.

Old Coastguard Station Visitor and Education Centre ❹

The Dock, Robin Hood's Bay, North Yorkshire
Tel: +44 (0)1947 885900
E: oldcoastguardstation@nationaltrust.org.uk
www.moors.uk.net

Located at the bottom of the bank in Robin Hood's Bay next to the slipway, the Old Coastguard Station centre features interactive models and displays about the bay and the seashore, and a model of the dinosaur tracks recently discovered on the coast. Throughout the summer holidays there are Dinosaur Coast Team activities and North East Geology Trust Dino Days.

Rotunda Museum ❺

Vernon Road, Scarborough
North Yorkshire YO11 2NN
Tel: +44 (0) 1723 367326
Email: museuminfo@scarborough.gov.uk
www.rotundamuseum.co.uk

The Rotunda Museum, overlooking the South Bay, was built in 1828 to a design suggested by William Smith, the 'Father of English Geology'. Re-opened in 2008, it is a jewel in the crown of Britain's geological heritage. The collection includes middle Jurassic fossil plants and a pristine Carboniferous plant collection.

Warrington Museum and Art Gallery ❻

Bold Street, Warrington
Cheshire WA1 1JB
Tel: +44 (0)1925 442733
Fax: +44 (0)1925 443257
www.warrington.gov.uk

Founded in 1848, the museum's Geology Gallery features 1000 rocks, fossils and minerals, and a model of the reptile *Ticinosuchus*. The museum organizes special events such as A Day with the Dinosaurs.

Whitby Museum, Library and Archive ❼
Pannett Park, Whitby
North Yorkshire YO21 1RE
Tel: +44 (0)1947 602908
E: keeper@whitbymuseum.org.uk
www.whitbymuseum.org.uk

This museum has a rich fossil collection, particularly the beautiful and gigantic marine reptiles. Also displayed are ammonites, nautiloids, belemnites, fish and plants, with over 200 'primary'-type specimens, mostly from lower and middle Jurassic strata.

World Museum Liverpool ❽
William Brown Street
Liverpool, L3 8EN
Tel: +44 (0)151 478 4393 (information desk)
www.liverpoolmuseums.org.uk

The museum's huge natural science collection includes an extensive range of fossils (40,000 specimens). The marine Ordovician, Silurian, Devonian and lower Carboniferous periods are particularly well represented. There is also a collection of fossils from the Pleistocene and Holocene epochs. Unusual items include an egg from the first discovered clutch of Oviraptor eggs, and a complete skeleton of the extinct Irish elk *Megaloceras giganteus*.

OUTDOORS

Formby, National Trust ❾
Victoria Road, Freshfield
Formby, Liverpool L37 1LJ
Tel: +44 (0)1704 878591
Fax: +44 (0)1704 835378
E: formby@nationaltrust.org.uk

This is a large area of beach, sand dunes and pine woods, 2 miles west of Formby, which is managed by the National Trust. The NT organizes special events and guided walks for visitors, and these are probably the best way to see the 4500-year-old footprints of red deer and ancient humans. The tracks are visible along a silt plain that runs the stretch of Formby beach near the dune front. The shifting tides make it difficult to cite precise locations; on some days the tracks may be covered up.

Hilbre Island LNR ❿
Dee Estuary (SSSI), Wirral–North Flintshire
Tel: +44 (0) 151 1632 4455 (Local Nature Reserve Ranger)
www.hilbreisland.org.uk

Hilbre, now a Local Nature Reserve, is the largest of three tidal islands at the tip of the Wirral Peninsula and is reached from Dee Lane slipway in West Kirby when the tide is out. Details of when you must leave Hilbre to get back safe and dry are clearly posted here. The uniqueness and clarity of the ancient reptile tracks found on Hilbre Island have led to its identification as a potential GCR site. Dates and information regarding events and activities arranged by the Wirral Ranger Service are available online at www.wirral.gov.uk. Permits to visit the island are available from the visitor centre (below, open daily 10.00–17.00).

Wirral Country Park and Visitor Centre ⓫
Station Road, Thurstaston
Wirral, Merseyside CH61 0HN
Tel: +44 (0)151 648 4371/3884
E: wirralcountrypark@wirral.gov.uk
www.wcpfg.org.uk

This was the first designated Country Park in Britain, based on the old West Kirby–Hooton branch line. Thurstaston is its main hub and the base of the Wirral Rangers. There is a visitor centre with an information desk for queries, including safe crossing times to Hilbre Island. Book here for the many ranger-led activities in the park.

Yorkshire Coast: 'Dinosaur Coast'
www.dinocoast.org.uk

The Jurassic Coast of North Yorkshire is one of the richest fossil resources in the country. The coast stretches for more than 40 miles, from Staithes in the north, through Whitby, Scarborough and Filey to Speeton in the south. The cliff and foreshore exposures comprise layers of rock laid down as marine and coastal plain sediments 190–130 million years ago. These rocks offer fossil-collectors a wide range of extremely well-preserved fossils. Common finds include ammonites and belemnites as well as marine reptiles and fish. Fossil-collecting is extremely popular, as countless fossils are eroded from the cliffs; unless collected, many would be lost to the sea. However, without proper care and attention, the scientific value of fossils can be damaged or destroyed.

YORKSHIRE COAST: FOSSIL HOTSPOTS

Robin Hood's Bay: This small fishing town, 5 miles south of Whitby, is famous for the fossils to be found on its beach.
Runswick Bay: On this rocky cliff coastline, the village of Runswick Bay, north of Whitby, is among the relatively few points at which it is possible to reach the shore. The large sandy beach is a great place to look for fossils such as ammonites and jet.
Scarborough: The history of Scarborough is bound up with the geology of the area. An abundance of specimens is to be found along the coastline.
Speeton: The substantially complete skeleton of a plesiosaur was found at Speeton by amateur collector Nigel Armstrong in 2001, who realized the importance of the find and called the local museums. It is now in the Rotunda Museum, Scarborough.

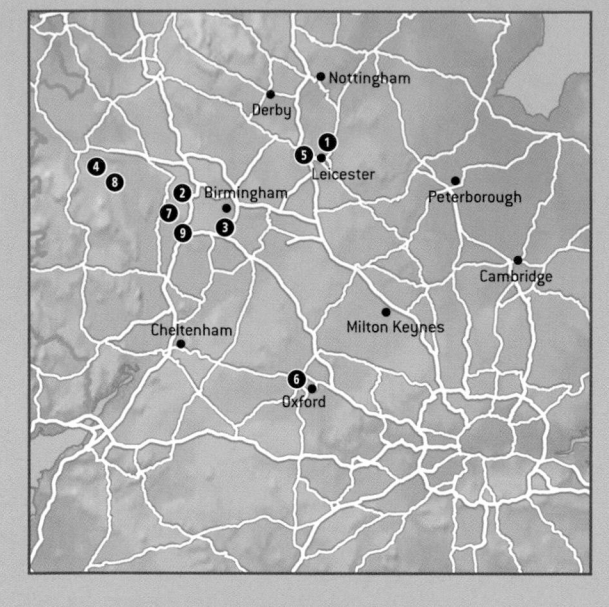

Central England

Charnwood Museum ❶

Queen's Hall, 1a Granby Sreet
Loughborough, Leicestershire LE11 3DU
Tel: +44 (0)1509 233754
E: charnwood@leics.gov.uk
www.leics.gov.uk

Situated in the city's Queen's Park, the museum offers a collection
of minerals, rocks and fossils found in Leicestershire and Rutland.
The geological collections are of national interest because of the
importance of the local Precambrian rocks. The fossils are mainly
from the Carboniferous and Jurassic periods. There is also a handling
collection of rocks, fossil and minerals for use by children and the
public in handling sessions, special events and school workshops.

Dudley Museum and Art Gallery ❷

St James Road, Dudley
West Midlands DY1 1HU
Tel: +44 (0)1384 815575
E: dudley.museum@dudley.gov.uk
www.dudley.gov.uk

This museum contains a huge collection of fossils for budding
geologists. The Fossil Gallery features a cross-section of the Silurian
and Carboniferous fossils of the region. Use the collection to identify
any fossils you found when visiting the Wren's Nest (see p. 183).

Lapworth Museum of Geology ❸

University of Birmingham
Edgbaston, Birmingham
West Midlands B15 2TT
Tel: +44 (0)121 414 7294/6751
Fax: +44 (0)121 414 4942
E: lapworth@contacts.bham.ac.uk

This museum has a fine collection of fossils, minerals and rocks, and
large collections of early geological maps, equipment, models and
zoological specimens. There are beautifully preserved fish,
dragonflies, crabs, lobsters and pterosaurs from the Solnhofen
Limestone in Germany, and outstanding fish collections from Brazil,
Italy, Lebanon and the USA. There are even some 510-million-year-old
Cambrian-period animals unique to the world-famous Burgess Shale
of British Columbia. Tours, talks and access to additional collections
are available by prior arrangement.

Much Wenlock Museum and Visitor Information Centre ❹

The Memorial Hall, High Street
Much Wenlock, Shropshire TF13 6HR
Tel: +44 (0)1952 727679
E: much.wenlock.museum@shropshire.gov.uk

The museum is located in the Memorial Hall at the lower end of the High
Street, opposite the town square; the visitor information centre is
located in the same building. The museum offers geological displays
about the area, including fossils.

New Walk Museum and Art Gallery ❺

New Walk, Leicester LE1 7EA
Tel: +44 (0)116 225 4900
E: museums@leicester.gov.uk

Situated within the historic New Walk area of the city, the museum
exhibits a wide variety of fossils and minerals. Leicester's oldest
fossil, *Charnia masoni*, is on display. The Mighty Dinosaurs display
enables you to walk in the footprints of giants and discover their
awesome power – and that of their maritime relatives, the
plesiosaurs and ichthyosaurs.

Oxford University Museum of Natural History ❻

Parks Road, Oxford OX1 3PW
Tel: +44 (1)0865 272950
Fax: +44 (1)0865 272970
E: info@oum.ox.ac.uk
www.oum.ox.ac.uk

Located on Parks Road, off Broad Street, central Oxford, the Museum
of Natural History offers a varied programme of temporary shows
plus a permanent exhibition of an impressive display of dinosaurs.
A long trackway of giant *Meglasaurus* prints displayed on the front

lawn sets the tone. Inside, there are four species of local dinosaur: *Eustreptospondylus*, *Camptosaurus*, *Megalosaurus* and *Cetiosaurus*, plus others from around the world, including *Iguanodon* and *Tyrannosaurus*. The dodo, another museum treasure, was immortalized by the author Lewis Carroll, a regular museum visitor, in *Alice's Adventures in Wonderland*.

OUTDOORS

Dudley Canal and Limestone Mines ❼
Duldley Canal Trust
501 Birmingham New Road
Dudley, West Midlands DY1 4SB
Tel: +44 (0)1384 236275
Fax: +44 (0)1384 456615
E: info@dudleycanaltrust.org.uk
www.dudleycanaltrust.org.uk

If you are interested in visiting the limestone mines at Dudley, a good way to see them is to take a canal boat trip into the heart of the caverns, which were carved during the search for raw limestone. The trip lasts 45 minutes and takes in five individual tunnels and the Little Tess and Singing Cavern mines, where audiovisual presentations retell the history of Dudley.

Wenlock Edge, National Trust ❽
Wenlock, Much Wenlock
Shropshire TF13 6DJ
Tel: National Trust +44 (0)1694 723068
Tel: Shropshire Wildlife Trust +44 (0)1743 284280
E: cardingmill@nationaltrust.org.uk
www.nationaltrust.org

This rare and special landscape stretches through Shropshire for some 15 miles. The thickly wooded limestone escarpment was laid down 420 million years ago and offers dramatic views, scenic paths, historic quarries as well as rare flowers, mammals, birds and insects. There are many fossils hidden in the area's rocks, including crinoids and trilobites.

Wren's Nest National Nature Reserve, Dudley ❾
Wren's Hill Road, Dudley, West Midlands
Tel: +44 (0)1384 812785 (warden)
www.dudley.gov.uk

A site of international importance with exceptional palaeontological features, located half a mile northeast of Dudley town centre. Visiting the reserve is a year-round outdoor activity. A team of wardens is available to help you explore the geology and many species of wildlife found at the park. Information on the area's geology and fossil identification is also available at Dudley Museum and Art Gallery (see left).

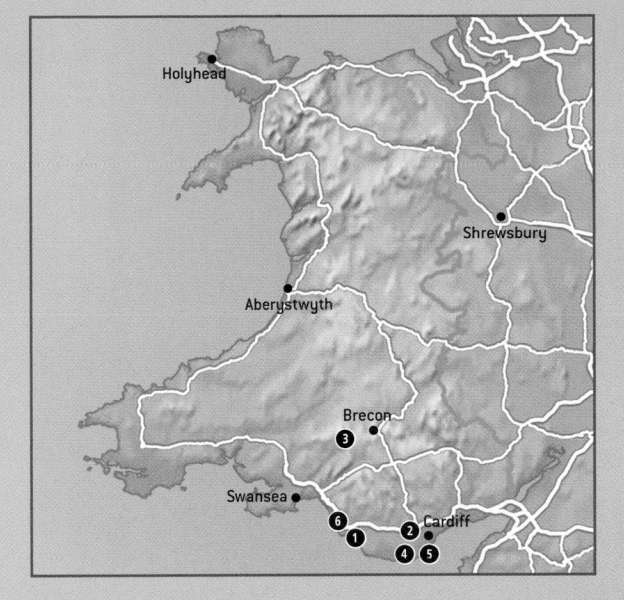

Wales

INDOORS

Glamorgan Heritage Coast Centre ❶
Dunraven Park, Southerndown
Vale of Glamorgan CF32 0RP
Tel: +44 (0)1656 880157

The centre offers a programme of themed walks and events, such as a fossils and geology day. Join the rangers on a walk from Southerdown beach to Ogmore to learn fascinating facts about the coast, its fossils and geology. The centre also organizes fossil-searching trips for school parties. Remember to check the tide times locally.

National Museum Cardiff ❷
Cathays Park, Cardiff CF10 3NP
Tel +44 (0)29 2057 3213
Fax +44 (0)29 2066 7332
E: geology@museumwales.ac.uk
www.museumwales.ac.uk

This is one of the National Museums of Wales, located in the Civic Centre. The Red Lady of Paviland is on display here, and the Evolution of Wales gallery features a large collection of fossils, notably brachiopods, trilobites, plants and dinosaur tracks. The exhibition covers the 4.6 billion years of Wales's history. The Glanely Discovery gallery is a hands-on space, where, with the help of specialist staff, you can find out more about fossils, get help identifying your finds, explore the handling collections and take part in activities.

National Showcaves Centre for Wales ❸

Glyntawe, near Abercrave
Brecon, Powys SA9 1GJ
Tel: +44 (0)1639 730 801 (24-hour information line)
E: info@showcaves.co.ukemail
www.showcaves.co.uk

Situated within the Brecon Beacons National Park, the award-winning centre offers a fascinating insight into cave formation via the exploration of three Carboniferous limestone caves: the Dan-yr-Ogof, Bone Cave and Cathedral Cave. The site also features one of the world's largest dinosaur parks: walk around over 50 life-size models; including the *Dimetredon*, *T. rex* and the graceful flying reptiles. There is also a themed Jurassic karting track for children. The site is a National Nature Reserve and offers free educational visits to Welsh schools.

OUTDOORS

Bendrick Rock ❹

Hayes Point, Sully, near Barry
Vale of Glamorgan, South Wales

Bendrick Rock can be reached via a path that follows the outside of the security fence round HMS *Cambria* at Hayes Point, Sully. This stretch of coastline and the rock headland of Bendrick Rock is one of Britain's most important areas for fossil tracks. It is an SSSI due to its geological features. For more information, see the 'Geological Walks in Wales' leaflets available from the South Wales Geologists Association:

South Wales Geologists Association ❺

c/o Dept of Geology
National Museum & Gallery Cardiff
Cathays Park, Cardiff CF10 3NP
Tel: +44 (0)29 2057 3213
E: geology@nmgw.ac.uk

Southerndown and Ogmore-by-Sea ❻

Vale of Glamorgan, South Wales

Located on the Glamorgan Heritage Coast, the sandy expanse of Dunraven Bay (Southerndown) is backed by dramatic cliffs. The area has the second-largest tidal range in the world. If you are planning a visit, it is advisable to call the Glamorgan Heritage Coast Centre to get tidal times. The distinctive cliffs surrounding Southerndown are 180–290 million years old and are packed with fossils. The layers are made up of limestone sandwiched by shale — a brittle substance that makes the cliffs unstable and dangerous. Visitors should be aware of the dangers of local extreme tides.

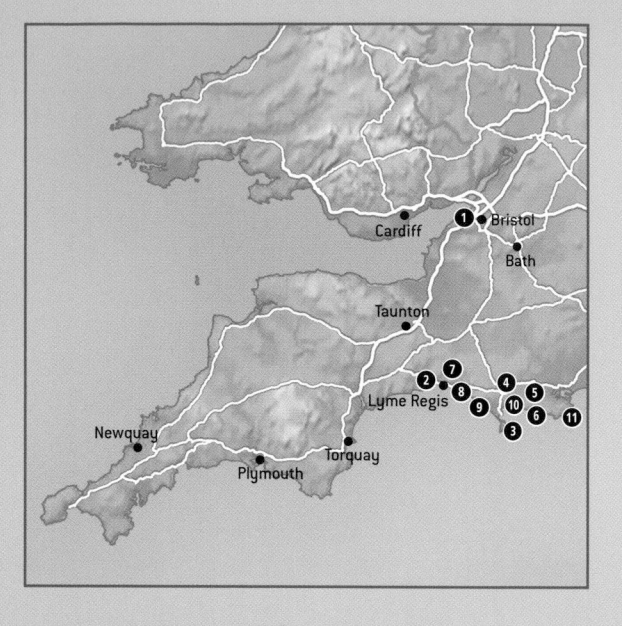

South-west England

INDOORS

Bristol's City Museum & Art Gallery ❶

Queen's Road, Bristol BS8 1RL
Tel: +44 (0)117 922 3571
Fax: +44 (0)117 922 2047
Minicom: 0117 922 3573
E: general.museum@bristol.gov.uk
www.bristol.gov.uk/museums

An extensive fossil and dinosaur collection can be found on the museum's first floor. The most complete dinosaur skeleton ever found in Britain, discovered in Charmouth, provides the new centrepiece for the popular dinosaur gallery, which has a fine collection of *Scelidosaur* fossils. The museum offers a dinosaur resource area for children. Its collection includes fossils from the Mendips and Triassic fossils of the Severn Estuary area.

Charmouth Heritage Coast Centre ❷

Lower Sea Lane, Charmouth
Bridport, Dorset DT6 6LL
Tel: +44 (0)1297 560772
E: info@charmouth.org
www.charmouth.org

This centre was set up to encourage safe and sustainable collecting of Jurassic fossils from the local beaches. The wardens here are able to give you advice on where to look, what you might find and how to use a hammer to find fossils. Fossil-hunting expeditions are organized throughout the year, as well as fossil weekends. Fossil roadshows are held on one day during most school holidays, with a year-round programme of events and activities for the public and schools. There are a number of ongoing research projects centred on local biodiversity, geology and coastal science. The centre has interactive computers, hands-on displays and information on fossils, fossil-hunting and the local coastal and marine wildlife, plus a video microscope to examine your finds. There is a Jurassic Theatre, where for a small charge (adults £1, children 50p) you can watch a short film on finding fossils at Charmouth or Black Ven. A Fossil Collecting Code, viewable via the website, has been devised by the centre.

Chesil Beach Visitor Centre ❸

Portland Beach Road, Portland, Dorset
Tel: +44 (0)1305 760579
Fax: +44 (0)1305 759692
E: reserve@chesil.fsworld.co.uk
www.chesilbeach.org

One of the Dorset Coastlink chain of visitor centres, the Chesil Beach Visitor Centre is located at the southern end of the Fleet Lagoon on the Ferrybridge car park between Weymouth and Portland, and provides information on this incredible natural area. Surrounding the centre is the Chesil Bank and the Fleet Nature Reserve, protected for wildlife and geology. Both are rare habitats and have been selected as a Special Protected Area (SPA) and a Special Area of Conservation (SAC). Inside the centre there is a touch table, an audio presentation, a touch-screen information point, a telescope and displays about wildlife, geology and history. A wide range of informative presentations and guided walks is offered during the year. The centre can be booked and illustrated talks arranged.

Dorset County Museum ❹

High West Street, Dorchester
Dorset DT1 1XA
Tel: +44 (0)1305 262735
Fax: +44 (0)1305 257180
E: enquiries@dorsetcountymuseum.org
www.dorsetcountymuseum.org

The museum's Jurassic Coast gallery, which opened in 2006, charts the 95 miles of Dorset and Devon coastline: how it was formed, what you can see today and why it is so important. This fascinating geological story is delivered through touch, sound, text and interactive displays, and at levels to suit all ages and all abilities. Children can follow the dinosaur tracks to measure themselves against a *Megalosaurus* and an enormous pliosaur flipper. The museum offers displays on past environments and the story of ammonites. Other displays include an ichthyosaur fossil and flying pterosaurs.

Kimmeridge Bay Marine Centre ❺

Kimmeridge Bay, Dorset
Tel: +44 (0)1929 481044

Managed by the Dorset Wildlife Trust, the centre explores the exceptionally rich marine life found on the numerous rocky ledges that run across the bay. It also interprets the wider marine environment found within the Purbeck Voluntary Marine Nature Reserve. 'Snorkelling for softies' and a live underwater camera are special features. Long rock ledges, a pebbled beach line and rocky outcrops all make the bay itself stunning. Kimmeridge boasts some of the most accessible marine wildlife in Dorset: an extended low tide provides extra rockpooling time, and the natural stone ledge jutting into the bay makes it easy to view the shallow waters.

Lulworth Heritage Centre ❻

Lulworth Cove, West Lulworth
Wareham, Dorset BH20 5RQ
Tel: +44 (0)1929 400587
Fax: +44 (0)1929 400155
E: estate.office@lulworth.com
www.lulworth.com/Education/heritage_centre.htm

The Lulworth Cove Heritage Centre has been enlarged and improved to form an interpretative exhibition of the countryside around the Lulworth area. Other parts of the exhibition include a fossil section together with displays depicting smuggling and sailing.

Lyme Regis Philpot Museum ❼

Bridge Street, Lyme Regis DT7 3QA
Tel: +44 (0)1297 443370
E: info@lymeregismuseum.co.uk
www.lymeregismuseum.co.uk

The museum is located in the heart of Lyme Regis, an area noted for its fossils. There is a fine display of rare fossils in the geological galleries. Guided fossil walks with the museum geologist are offered. The ticket price includes free museum access. You can also learn all about Mary Anning, the great Victorian fossil-hunter, who once lived on the site of the museum.

OUTDOORS

Dorset: Jurassic Coast

The 95-mile stretch of coastline from Orcombe Point (near Exmouth in Devon) to Old Harry Rocks (near Studland in Dorset) has an impressive fossil heritage and World Heritage Site status. Its boundaries are limited to the coastal strip that includes the current cliff tops and runs down to the level of the lowest tide. Because of site sensitivities, fossil-collecting is only recommended at Charmouth and Lyme Regis.

Jurassic Coast: Fossil-Collecting Code

1 It is best to go fossil detecting after a period of heavy storms as this is when the waves erode the cliffs, churn up the beach and reveal the fossils.

2 Eyes rather than hammers are your best tools. The best place to find fossils is on the beach.

3 It is very dangerous, and not a good idea, to dig for fossils in cliffs or landslides.

4 Watch out for the tides. It is easy to get cut off on some beaches, particularly on the Jurassic Coast because of the huge landslip blocking the beach between Lyme Regis and Charmouth, which is one of the best places to go fossil-detecting.

5 If your fossil is caked in clay and mud, soak it in a bucket for a few weeks and clean off the sediment gently with an old toothbrush.

6 If you are not sure what your fossil might be, take it to a museum, heritage centre or fossil roadshow event to get it identified. The Jurassic Coast hosts a fossil roadshow with the Natural History Museum every other year in Lyme Regis.

7 Fossil walks, events and activities are advertised on the virtual What's On guide at www.jurassiccoast.com.

8 The best way to get started is to join a guided fossil-collecting walk from local visitor centres or museums.

DORSET JURASSIC COAST: FOSSIL HOTSPOTS

Charmouth: A gateway town to the Jurassic Coast, Charmouth lies in the southwest corner of Dorset at the heart of the Lyme Bay coastline. The cliffs at Charmouth are part of the West Dorset Coast SSSI and are renowned for the range and quality of their fossils. A cast of a fossil *Scelidosaurus*, Charmouth's very own local dinosaur found in 2000, is sited on the wall of the Heritage Coast Centre.

Lyme Regis: The town is nicknamed 'Fossil Town' because it is a key centre for fossil-hunting. Massive natural erosion in the area is constantly exposing new crops of fossils. The Undercliff, stretching between the cliffs and the sea for 21 miles to the west of Lyme, is an unspoilt tree-clad wilderness sheltering such a wide diversity of plants and animals that in 1959 it became one of the first of Britain's National Nature Reserves. The National Oceanographic Centre at Southampton University keeps an extensive website on Lyme's geology, at www.soton.ac.uk/ffimw/lyme.htm.

Lyme Regis Fossil Festival ❽

Lyme Regis Development Trust
Unit 1, St Michael's Business Centre
Church Street, Lyme Regis
Dorset DT7 3DB
E: info@lymeregisfossilfestival.co.uk
www.lrdt.co.uk

The annual fossil festival features walks, talks and workshops, with palaeontologists from the Natural History Museum in London helping to identify fossil finds and run activities. The festival also celebrates the life of Mary Anning and the town's unparalleled role in the development of the earth sciences.

Dorset: Learning about Jurassic Coast Fossils and Geology

Burton Bradstock ❾

West Dorset
www.burtonbradstock.org.uk.

Burton Bradstock is at the westerly end of Chesil Beach, nestling in the Bride valley. See the website for a description of the Jurassic formations around Burton Bradstock and the structure of Chesil Beach.

Lulworth Cove ❿

Near West Lulworth, Dorset

Lulworth Cove is the most enclosed and sheltered harbour along this part of the coast. This is where Purbeck Marble rests on Portland Stone. Purbeck Marble (late Jurassic–early Cretaceous) is not a true marble, which is a metamorphosed limestone, but a lithified freshwater limestone that will take a good polish. It is rich in the freshwater snail *Viviparus* and was much used during the Middle Ages as a distinctive decorative stone in cathedrals, particularly in the south. On the east side of Lulworth Cove is a spot named the Fossil Forest where you can see the stumps of giant prehistoric trees that have turned to stone.

Durlston Country Park and Visitor Centre ⓫

Lighthouse Road, Swanage
Dorset BH19 2JL
Tel/Fax: +44 (0)1929 424443
E: info@durlston.co.uk
www.durlston.co.uk

On the southeast tip of Purbeck, about 1 mile from Swanage, this large National Nature Reserve features sea cliffs, coastal limestone downland, haymeadows, hedgerows and woodland. The reserve is the place to go for great views, walks, superb geology and a fascinating array of wildlife. It is a gateway to the Jurassic Coast World Heritage Site. The visitor centre has monthly displays, guided tours and full information on the area.

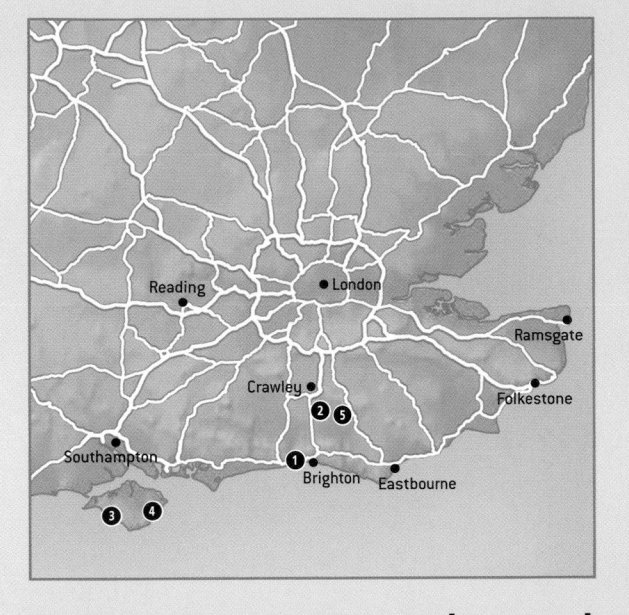

The Isle of Wight and Southeast England

INDOORS

The Booth Museum of Natural History ❶

194 Dyke Road, Brighton
East Sussex BN1 5AA
Tel +44 (0)1273 292 777
E: boothmuseum@brighton-hove.gov.uk
www.virtualmuseum.info

More than half a million specimens are housed in this fascinating museum. These include whale and dinosaur bones. The geology collection includes important insects in amber, and examples of rocks and minerals. The library includes over 14,000 natural history texts dating from the seventeenth century to the present day. Children can explore the collections through interactive displays in a new hands-on gallery. There is a regular programme of temporary exhibitions and events for adults, children and schools.

Cuckfield Museum ❷

Queen's Hall, High Street
Cuckfield, West Sussex RH17 5EL
E: cuckfieldmuseum@btconnect.com
www.cuckfield.org

Located on the first floor of the Queen's Hall, this local history museum tells a story unique to Cuckfield — the discovery of the *Iguanodon* by Dr Gideon A. Mantell and his wife. The museum display, with fossils of the Cuckfield dinosaur, and a cast and a replica of

dinosaur tracks found on Bexhill Beach, complements the memorial to Mantell at nearby Whiteman's Green. There are bones, photographs, archive and documentary material relating to this discovery and to the general geology of the area.

Dinosaur Farm Museum ❸

Military Road, Brighstone
Isle of Wight PO30 4PG
Tel: +44 (0)1983 740844
www.dinosaurfarm.co.uk

This museum is situated on the Military Road (coast road A3055) near the village of Brighstone on the southwest coast. Established in 1993 after the discovery of a brachiosaur-like dinosaur known as the Barnes High Sauropod, the museum has three renovated rooms for display and is staffed primarily by volunteers who clean the dinosaur bones found on the Isle of Wight's beaches. All the dinosaurs are from the earliest part of the Cretaceous period (145–65 million years ago); the oldest dinosaur bones are 132 million years old and the youngest 110 million years old. Children's activities include 3D dinosaur jigsaws, a fossil sandpit and dinosaur rubbings. A variety of fossils is on display, and these change regularly as more are discovered. The museum offers a free fossil identification service and guided fossil-hunts throughout the year. Parties of up to 25 are taken on a guided tour of a local beach.

Dinosaur Isle, Isle of Wight ❹

Culver Parade, Sandown
Isle of Wight PO36 8QA
Tel: +44 (0)1983 404344
Fax: +44 (0)1983 407502
E: dinosaur@iow.gov.uk
www.dinosaurisle.com

This dinosaur museum is situated on the seafront at Sandown Bay. About 30,000 geological specimens, mainly fossils, are housed here. A large dinosaur gallery has lifelike reconstructions. There is also a wide range of activities, including talks and fossil-handling sessions, guided visits and field trips, such as visits to local coastal sites. An identification service for Isle of Wight fossils is also offered, and there is an education room for talks and activities.

OUTDOORS

The Mantell Monument ❺

Whiteman's Green
Near Cuckfield, West Sussex

The monument marks the site of the former stone quarry on the northern edge of the village where Mantell discovered remains of the dinosaurs *Megalosaurus, Iguanodon, Hylaeosaurus* and *Pelosaurus* during the years 1817–47.

SOUTHEAST COAST CHALK: FOSSIL HOTSPOTS

(SSSI restrictions apply)

Bracklesham Bay, West Sussex: A good variety of fossils may be found here, including shells, sharks' teeth, turtle shell and corals.

Eastbourne, East Sussex: This area is well known for the high quality and size of its fossils, including ammonites, nautilus and echinoids.

Folkstone, Kent: This is one of the best places to find fossils on the south coast. The Gault Cliffs are rapidly eroding to reveal ammonites, belemnites, echinoids and more. Fossils here tend to be fragile. Most fossil finds can be discovered not only on the foreshore, especially after storms, but also in the cliffs.

Peacehaven, East Sussex: This is the best place to view some of the world's largest and most spectacular ammonites, but fossil-collecting is prohibited. Friar's Bay, between Newhaven and Peacehaven, is a good viewing location. Certain areas do permit fossil-collecting of other marine fossils that lived alongside ammonites.

Seven Sisters, East Sussex: This area is most famous for a wide variety of good quality Cretaceous fossils.

ISLE OF WIGHT: FOSSIL HOTSPOTS

Compton Bay: This sandy beach is located beside the coast road between Freshwater Bay and Brook. It is an area popular with fossil-hunters. At low tide, dinosaur tracks can be seen imprinted in the rocks.

Hanover Point: This stretch of beach, from Compton Chine to Brook Chine, is easily accessed from the car parks located at either end. The site is protected by the National Trust. The giant casts of dinosaur tracks in stone are a famous feature.

St Catherine's Point: This beach at the southern tip of the island is composed of sandstone, chalk and chert boulders, and is rich in fossils, including ammonites and shells. This is an outstandingly beautiful area and is accessible even at high tide. There is a walk downhill to the beach.

Shanklin Beach: This beach is only accessible on a falling tide. Sandstone boulders on the beach, which forms part of Sandown Bay, contain fossilized shells. Fossilized wood is also commonly found here.

Yaverland Beach: The coast between Yaverland and Culver Cliff forms the northern end of Sandown Bay. This sandy beach is easy to walk on, but access is via a short set of steps. It is about 1 mile from Sandown. The beach is inaccessible at high tide.

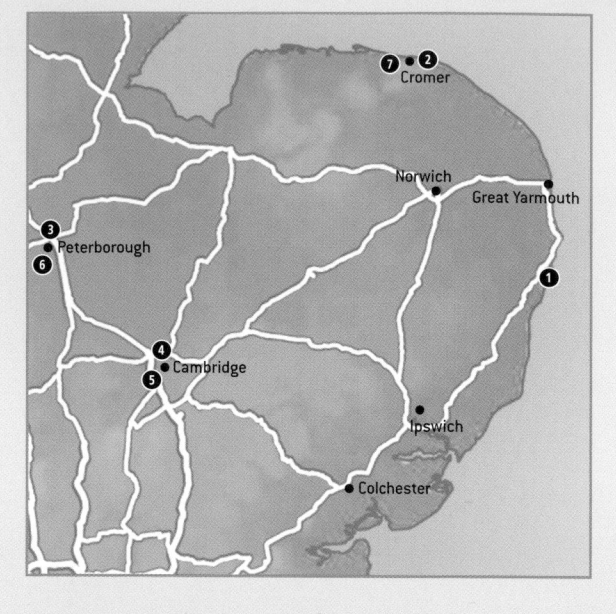

The East of England

INDOORS

The Amber Museum and Shop ❶
15 Market Place, Southwold
Suffolk IP18 6EA
Tel: +44 (0)1502 723394
Fax: +44 (0)1502 723929
www.ambershop.co.uk

This is a privately run museum that tells the story and history of the amber that is found on the Suffolk coast, and offers pieces for sale. The museum features displays of amber artefacts, carvings, jewellery and objets d'art, modern and antique. It also houses many large examples of amber found on the English coast, including the largest at 2.2 kg.

Cromer Museum ❷
East Cottages, Tucker Street
Cromer, Norfolk NR27 9HB
Tel: +44 (0)1263 513543
E: cromer.museum@norfolk.gov.uk

The museum is situated near the tall tower of Cromer Church in the centre of town in a Victorian fisherman's cottage. A geology gallery features a fine collection of fossils – all found in Norfolk – and displays that reveal why Cromer is renowned as a geological area of international importance. Highlights include the West Runton Elephant, Britain's oldest and most complete elephant fossil, plus actual bones; and a cast of the skull of a mosasaur, a huge marine reptile common off the north Norfolk coast about 80 million years ago. A year-round events programme includes guided geology walks to look at rocks and fossils, and school holiday fossil-handling.

Peterborough Museum ❸

Priestgate, Peterborough
Cambridgeshire PE1 1LF
Tel: +44 (0)1733 343329
Fax: +44 (0)1733 341928
E: museum@peterborough.gov.uk
www.peterboroughheritage.org.uk

The museum is situated in the city centre just off the main shopping area – follow the pedestrian signs. A highlight is the internationally renowned collection of marine reptile fossils, collected locally from Jurassic Oxford Clay beds. It includes the world's most complete plesiosaur, plus the largest fish that ever lived – *Leedsichthys* – which can be viewed by appointment only. Other Jurassic marine fossils include ammonites and belemnites. There are also Ice Age remains of mammoths, giant reindeer and a hippopotamus.

Sedgwick Museum of Earth Sciences ❹

Downing Street, Cambridge CB2 3EQ
Tel: +44 (0)1223 333456
Fax: +44 (0)1223 333450
E: sedgwickmuseum@esc.cam.ac.uk
www.sedgwickmuseum.org

The Sedgwick Museum has fossils, rocks and minerals from around the world. A complete *Iguanodon* skeleton forms the centrepiece near the entrance. Its ancient life section features the world's largest spider, desert creatures, and an exploration display of the Wenlock Reef featuring some 3500 fossils. Life in the Jurassic seas is represented by large marine predators, ammonites, fish and echinoids. Other dinosaur highlights include a *Megalosaurus* skull and reconstructions of *Compsognathus* and *Velociraptor*. There are displays on the origins of modern life, 213 million years ago to the present day, with starfish, ammonites, *Archaeopteryx*, giant crocodiles, coprolites, turtles, fish, plants and whale ear bones.

University Museum of Zoology, Cambridge ❺

Downing Street, Cambridge CB2 3EJ
Tel: +44 (0)1223 336650
Fax: +44 (0)1223 336679
E: umzc@zoo.cam.ac.uk
www.zoo.cam.ac.uk

Situated in central Cambridge on the University New Museum site, entry to the museum is through the archway on Downing Street, opposite Tennis Court Road. As you enter you will notice the finback whale skeleton hung above the entrance. Inside are large, modern display galleries and a collection of outstanding historical and international importance: much of the museum's material is from the great collecting expeditions of the nineteenth century, which provided the first documentation of faunas in many parts of the globe. They include individual specimens of exceptional historical significance, including fine examples of the dodo and great auk, skins of the extinct Tasmanian wolf, and many of Darwin's specimens, some collected from his voyage on the *Beagle* and from his time studying at Cambridge. Auroch skeletons are on display in the Lower Gallery. New arrivals include fossils of the earliest land vertebrates, molluscs from the excavation of the Channel Tunnel that document climatic change in Europe over the past 10,000 years, and a rich variety of invertebrate fauna from the Seychelles.

OUTDOORS

King's Dyke Brick Pit ❻

Whittlesey, near Peterborough
Cambridgeshire

Peterborough and the surrounding area have seven designated RIG (Regionally Important Geological or Geomorphological) sites. RIG sites, or RIGs, have been evaluated for their scientific, educational, historic, aesthetic and conservation value as well as their access and safety. The site at King's Dyke Brick Pit has been developed and opened to the public to provide a place for a wide range of people to come and enjoy fossil-hunting and to become more aware of the geological heritage of Whittlesey and Peterborough. There are on-site interpretation boards that provide general information about the site.

West Runton ❼

West Runton, near Cromer, Norfolk

The fossil-filled West Runton Freshwater Bed is just east of West Runton on the north Norfolk coast. The area is part of the Cromer Ridge, which was created by ice sheets melting at the end of the last Ice Age. The rock pools on West Runton beach are a must for fossil-hunters. A search following a stormy day and a high tide is the best time for fossil-collectors, as the area will expose yet another crop of fossils.

West Runton Geology Walks

Tel: +44 (0)1263 513543 (Cromer Museum)
www.museums.norfolk.gov.uk

Join Cromer Museum's curator to look at the rocks and fossils of West Runton Beach. Examine the Ice Age structures in the cliff and visit the site of the famous West Runton elephant excavation that reveals so much about life in Norfolk over half a million years ago. If you have any fossils that need identifying, take those along too. Booking is essential.

Index

Acknowledgements

The authors would like to thank the many people who helped with this book, especially the BBC team who produced the television series: Kerensa Jennings, Gavin Boyland, Amanda Kear, Andy Hawley, Toby Strong, Mark McCauley, Katie Elloway, Fiona Pitcher and all the contributors to the programmes, including David Attenborough, Mike Bassett, Matthew Bennett, Alan Bowden, John Carney, Neil Clark, Sue Cooke, Joe Crossley, Robin Fournel, Nick Hogg, Anjana Khatwa, Nigel Larkin, Hilton Law, Jeff Liston, Phil Manning, Mike Marshall, Gordon Roberts, Eric Robinson, Adrian Shine, Martin Simpson, Derek Siveter, Geoff Tresise, Nigel Trewin, Phil Wilby and James Wong.

We would also like to thank the Open University, in particular Peter Sheldon for his expertise and enthusiastic support, Janet Sumner and Alison Ramsey; Georgina Capel at Capel and Land Ltd; Colin Scrutton; and all those at Tall Tree and BBC Books involved with *Fossil Detectives*. Hermione would especially like to thank Jon Peltenburg for all his help.